INTUITION

INTUITION

Unlock Your Brain's Potential
To Build Real Intuition

JOEL PEARSON

WELBECK

Copyright © 2024 Joel Pearson

The right of Joel Pearson to be identified as the Authors of
the Work has been asserted by him in accordance with the
Copyright, Designs and Patents Act 1988.

First published in 2024 by Welbeck Balance
This edition published by Welbeck
An imprint of Headline Publishing Group Limited

SRD

Apart from any use permitted under UK copyright law, this publication
may only be reproduced, stored, or transmitted, in any form, or by any
means, with prior permission in writing of the publishers or, in the case
of reprographic production, in accordance with the terms of licences
issued by the Copyright Licensing Agency.

Cataloguing in Publication Data is available from the British Library

ISBN 978-1-80129-306-8

Printed and bound in India by Manipal Technologies Limited, Manipal

MIX
Paper | Supporting
responsible forestry
FSC™ C104740

Headline's policy is to use papers that are natural, renewable and recyclable
products and made from wood grown in well-managed forests and other
controlled sources. The logging and manufacturing processes are expected
to conform to the environmental regulations of the country of origin.

HEADLINE PUBLISHING GROUP LIMITED
An Hachette UK Company
Carmelite House
50 Victoria Embankment
London EC4Y 0DZ

The authorised representative in the EEA is Hachette Ireland,
8 Castlecourt Centre, Dublin 15, D15 XTP3, Ireland (email: info@hbgi.ie)

www.headline.co.uk
www.hachette.co.uk

'*Intuition* offers a breakthrough perspective on intuition based on neurological research. I loved the clarity in its definition of intuition; it is inspiring and deeply original, transforming abstract concepts into a practical guide for developing and harnessing intuition in our lives.'

Ed Catmull, *New York Times* bestselling author of *Creativity, Inc.* and co-founder of Pixar

'A superb guide to the science and art of intuition. Pearson marries the science of psychology with a string of fascinating anecdotes to explain when and why humans should rely on intuition, and wraps his ideas in a framework that's as useful as it is memorable. Highly recommended!'

Adam Alter, *New York Times* bestselling author of *Irresistible* and *Anatomy of a Breakthrough*

ABOUT THE AUTHOR

Joel Pearson is a psychologist as well as a neuroscientist, and is the Director of Future Minds Lab at the University of New South Wales. He is also Head of Innovation & Enterprise in the School of Psychology at the same institution. He initially studied art and filmmaking at the College of Fine Arts, UNSW, before turning to the scientific mysteries of human consciousness and the complexities of the brain. His pioneering research has changed our understanding of both intuition and the human imagination. He is a prolific public speaker and writer, having contributed chapters to several books and published numerous journal articles. Joel's website is at profjoelpearson.com.

Contents

PART TWO

THE FIVE RULES FOR INTUITION

PART THREE

THE PRACTICE OF INTUITION

Introduction

Intuition in action: dodging disaster in the skies

On a clear day with the kind of expansive deep blue sky that only Australia seems to offer, Jason sat comfortably in the cockpit of his A-4 Skyhawk flying a practice session for an upcoming air show. The Skyhawk is an older plane but still a proper fighter jet: agile, loud, all metal, capable of extreme warfare, the type you typically see only in movies.

He and another pilot were flying formation near the town of Nowra, south of Sydney.

Jason is a clean-cut, well-groomed man, precise in his definitions, descriptions and actions, in the way military personnel often are. The straining and screaming of the jet engines was a familiar, almost comforting sound to him. When flying formation, it's the job of the pilot who's following in the rear – in this case, Jason – to stay 100 per cent focused on the lead plane and to mirror its moves exactly: ascents, descents, turns or rolls. So that from the ground, the two planes appear to move in perfect synchrony.

In this particular session, Jason and the lead pilot were practising several manoeuvres, including one called a barrel roll. In a barrel roll the aircraft rotates around its own axis, spinning as it flies, whilst at the same time flying in large loops – a bit like spiralling around the inside of a giant barrel. Everything was going as it should: both pilots felt good, they were well rested, the sky was clear, and the planes were flying perfectly.

Then something went tragically wrong. As the two aircraft began to pull out of the roll, it suddenly became clear that they were flying too close to the ground. The lead pilot sounded an emergency call: 'Pull up, pull up!'

Introduction

But Jason found that he had in fact already pulled back on the plane's joystick, well before the lead pilot called the emergency. Keep in mind that Jason's job when flying formation was to wait for the lead aircraft to manoeuvre and then mirror those moves. Yet somehow, without knowing why or when, he had done the opposite; his body had begun reacting to the situation before the lead pilot's call, before even his own mind had grasped what was happening.

Data from the aircraft later confirmed that Jason had begun to pull his aircraft up before the emergency call. Tragically, the lead pilot did not react in time; his plane crashed and he was killed on impact.

Jason's precognition response in this scenario is amazing, particularly given that he was flying formation. Pilots are taught right from the start to use their instruments, and not to rely on feeling, bodily sensations, or what they see out the window. There are many stories of pilots flying out of clouds upside down: unable to believe what their instruments are telling them, they keep adjusting the aircraft based on their own, inaccurate sensory information, until they are completely inverted.

Despite, or perhaps because of, Jason's extensive training and his decade of experience flying fighter jets, all the sensory cues that his brain was processing during that practice session, some conscious, some not, brought his brain to the conclusion that something was not right. His brain instructed him to act without having to think through the process consciously. He responded to his brain's call for action, saving his aircraft and himself.

In the crucial seconds or milliseconds between the end of the manoeuvre and the lead pilot's impact, Jason's brain had processed all the real-time sensory information, too rich and fast for him to be fully conscious of, which then triggered negative associations in his brain, based on thousands of hours of experience in similar contexts. The negative sensation immediately transferred to action signals in Jason's motor cortex, and he pulled back on the joystick, bringing his aircraft's nose up just in time.

This is intuition in action. Against all the odds, the brain of a pilot trained to use the plane's instruments, and to focus only on the lead plane, was nevertheless tracking everything going on, unconsciously processing a mass of information from all his senses, crunching the data and rapidly concluding that something was amiss, until his arms pulled back on the joystick.

Jason's example cuts to my definition of intuition:

Introduction

Intuition is the learnt, productive use of unconscious information to improve decisions or actions. Jason's brain was processing many streams of incoming information in real time, and because he'd flown so often before, the information was already associated with positive or negative outcomes. These associations translated into brief feelings – *gut feelings* – that rapidly drove Jason to the action of pulling back on the throttle.

Jason's story sums up what this book is about. It strives to make sense of the ability to tap into the extra information normally hidden outside of awareness, and to provide a new, all-encompassing theory and practical guide for everyone to understand intuition and learn to use it safely.

But how does this example of a pilot in a life-and-death situation relate to you and me, everyday people on a quest to becoming better decision-makers? There are many extreme cases of intuition saving lives, but the same principles apply when you choose a café for lunch, say, or decide to trust someone you've just met, or to go on that second date. It applies when you, as a parent, know that something is up with one of your kids; and every time you make a choice or perform a rapid action in sport, or in your car, or at work, or in countless other day-to-day scenarios. You have the opportunity to tap into a vast sea of unconscious information, sourced from your five senses, often sitting there unused. With training, this extra data can enable better decisions and actions in all of us.

We've all heard the saying 'go with your gut'; we may have asked someone facing a tricky decision, 'What is your intuition telling you?' Maybe we can think of a time when we acted intuitively, without consciously or logically processing our surroundings. This book explores the layers of these intuitive processes. It breaks down and analyses what's actually happening in your brain when you tap into your intuition to make a decision or take action. It simmers down all the neuroscience and psychology to demystify intuition and provide a safe and reliable everyday guide to your decisions, in the form of five easy rules.

I have spent the past ten years actively researching human intuition, and my lab and I were the first to develop a scientific test to measure intuition. Before this, scientists weren't even agreed that intuition was a real thing, or they argued about how to define it. I am now on a mission to get the new science of intuition out into the world, where it can have a massive transformative impact.

A lot of people already use intuition, sometimes on a daily basis. Partially its use is ingrained – that moment when you go quiet and check in with how your body is feeling in order to decide something – but it's also something many have learnt to rely on, particularly in decision-making. Importantly, as we will see, there is such a thing as the misuse of intuition, so how can we know if we're using it in the right way? How can we be sure that the decisions and actions that feel intuitive are not based on misunderstandings or biases fooling us into making bad choices?

Because a lot of the science of intuition is new, or hidden behind another name, the information that *is* available is often not backed up by scientific evidence. Intuition has even become wrapped up with woo-woo wellness and spirituality, and sometimes gets defined as a magical sixth sense, outside the realm of science. In this book, I sift through all the most up-to-date science of intuition, consciousness, learning and decision-making, and transform it into five science-based rules that allow the safe and informed use of intuition.

Our brains collect and store immense amounts of data from the world around us that we are completely oblivious to – it remains unconscious. The five rules of intuition allow us to tap into that great reservoir of information, giving us a clear advantage for better, faster, more confident decisions – from the boardroom to the sports field, from a pilot to a busy parent, from complex life-changing decisions to the most simple. But let's be clear: there are also many situations when we should *not* use intuition, and this book lays out exactly how to distinguish those. My science-first approach mainstreams intuition, through compelling real-life stories to build an evidence-based practice.

A mission statement

This book does not set out to be an in-depth textbook on intuition: its mission is to distil the science of intuition into simple, practical rules that are easy to follow and can improve decision-making. I have taken the knowledge gained from the research done in my lab, and also drawn on the research of many others, in other disciplines and on other related topics, to help you to understand when you can safely use your intuition.

Introduction

The science of intuition is in its infancy, and there is far more yet to discover than is already known. However, this does not prevent the application of the knowledge we already have.

Psychologists have argued a lot about whether intuition is good and useful or dangerous and to be avoided. Indeed, there are many examples in books and in scientific literature of terrible cases where intuition has led people astray, but, as we shall see, these are in fact tragic cases of something I call misintuition. There are also many stories of intuitive feelings saving lives. The fact is that no one blanket statement – 'intuition is good' or 'intuition is bad' – is either truthful or practically useful once you understand how the human brain and mind work. If you know when you should and should not use intuition, you can reduce the negative potential, and avoid the situations where it will lead you astray while maximising its benefits.

The definition for intuition that I will use in this book is *the learnt productive use of unconscious information to improve decisions or actions*. Today, in business, for example, most leaders have to make rapid decisions with only a small amount of conscious information available, and intuitive decision-making plays a key role. Indeed, the speed at which things are changing is now faster than ever before and uncertainty is growing, therefore the need for intuition in business is only going to grow along with it. But here's the catch: how many people in leadership positions feel comfortable disclosing that they use intuition? How many would tell their employees, their board of directors, or the public for that matter, that they made an important decision using their intuition? My hope is that by spreading the word about the science of intuition, those who use it, even at the highest levels of decision-making, will be more comfortable discussing it.

By creating a science of intuition, we allow people to understand and tap into a real, measurable human ability. An extraordinary gift of the brain. Intuition can be explained by phenomena we already know from neuro-science and psychology. We don't need to bring in ideas about something extrasensory. Or ideas of collective intelligence, collective unconscious, or indeed anything magical or spiritual. Intuition is real. Science can explain it.

The five rules for intuition

Drawing from extensive studies in psychology and neuro-science conducted over many years, these rules provide a framework for understanding and safely using your intuition. For ease of memory, I have summed up the rules with the acronym SMILE:

S Self-awareness (Feeling emotional? Don't trust your intuition)

M Mastery (Learn before you leap: why mastery matters for intuition)

I Impulses and addiction (Never mistake impulsive desires for intuitive insights)

L Low probability (Resist the temptation to use intuition for probabilistic judgements)

E Environment (Only use intuition in familiar and predictable contexts)

Each science-backed rule is illustrated with stories of real-world, life-changing examples of intuition. You'll meet an intuitive mountain climber, learn about the gameshow that uses our innate misunderstanding of probability to rig its prizes, discover how the movie *Inception* inspired my lab's breakthrough on intuition science, learn why you should never go rock-climbing on a first date, and discover what happens when Usain Bolt races in low gravity.

You will learn why self-awareness is so important and why you shouldn't trust your intuition when you're emotional. And why you need mastery in an area before you can trust your intuition for it. Intuition is learnt, and you need to develop it for each area of your life.

Think about the last time you made a decision based on intuition. Did you consider your emotional state? Your prior experience in similar situations? Were any primal brain impulses, such as a craving or an addiction, leading you astray? Or did you make a bad choice based on the probabilities in play? Did you make sure you were in a familiar environment? In other words, were you able to use the untapped power of your unconscious brain to make the best decision available to you? No? Then this book will teach you how. But first, let's look more closely at the power of intuition, why it's a real thing, and the new science of measuring it.

PART ONE

The Background

The Power of Intuition

Feeling the unconscious

Jon Muir's story is one of incredible passion, relentless determination, and unparalleled adventure. Born just outside the Australian coastal town of Wollongong in 1961, Muir was just fourteen when his life was forever changed by a documentary about Mount Everest. From that moment on he knew that his destiny lay in the mountains. Jon has many stories of being saved by his gut-based decisions, whether it be while solo kayaking or climbing or trekking across deserts. 'The difference between life and death in extreme adventures,' he says, 'often comes down to intuitive decisions.'

Muir began his training to become a professional rock climber as a teenager, scaling the local cliffs around Illawarra's iconic escarpment, just south of Sydney. His hunger for adventure led him to make the bold decision to drop out of school and pursue his dream full-time. He has since travelled and climbed all over the world, and in 1988 became the first person to climb Mount Everest from the south side without a Sherpa. He has kayaked oceans, trekked the North Pole, and walked across Australia unassisted, setting numerous world records.

On the day I meet him, Jon doesn't look like your typical adventurer. He is not clad in high-tech, extreme outdoors gear, but dressed in an outfit of a singlet, kilt, huge hiking boots, and, under his beard, layers of leather necklaces sporting a Noah's Ark-worthy range of different animal teeth. His eyes are alive with adventure, they sparkle with the unexpected, and his voice is genuine and unfiltered, mainlining his personal way of talking. We're talking about the attempt Jon and a team made to summit Everest in 1984.

'In those days you couldn't just buy your way to the summit,' he says with a laugh. At that time, the mountain had been climbed fewer than a hundred times, and a summit attempt was not to be taken lightly. It was years in the making; the amount of preparation, travel and focus required was immense, putting all the more pressure on pushing all the way to the summit when the final day came.

Five team members set out to make the summit that morning; one stayed behind at the camp, feeling too tired to go. In the early hours of the morning it was still dark and bitterly cold. The wind howled as it contoured over the world's most famous mountain. The team had camped for the night – if you can call only a few hours of sleep 'the night' – about 8,000 metres up on the west ridge, which had been climbed only once or twice before.

It wasn't like you see nowadays on Netflix or the Discovery channel, with huge teams of Sherpas carrying all the bags, ropes, ladders and oxygen tanks, and climbers being supervised by safety officers all the way. In 1984 there was none of that. No high-tech clothes or energy gels, just six people packing up a tent in the bitter cold dark.

Moving slowly up a steep gully, the five were initially shel-
tered from the icy wind that was ripping up the mountain at
100 miles per hour, but Jon was worried all morning about
that wind. He knew that if it dropped they would be okay,
they could reach the summit; if not, there could be trouble.

Jon is famous in the adventurer community for having
a stainless-steel stomach. He never gets sick, even eating at
street stalls from India to Thailand – nothing sets off his
stomach. But on this particular morning his stomach was
heavy. It was an odd feeling of his stomach sinking and
pulling him down. A feeling of dis-ease, as he describes it.

At first, he didn't know why or what was causing it.
With each small slow step up the mountain, the feeling got
stronger; it started to feel like a little voice inside him saying
things might not be okay. It was late summer, which was
late in the Everest season, it being too cold to summit in the
winter, and this was their one shot. It was all or nothing – if
they called it off now, that would be it for the year. At these
heights, the air is thin, dangerously low in oxygen, which
clouds your judgement. 'The momentum was all up up up,'
says Jon.

And then Jon stopped dead still on the mountain. 'This
isn't right,' he told the team, 'this is all wrong. It's going to

be too windy up there. We're going to get frostbite at the very best or blown off the mountain at worst.'

The others stopped in shock, offended that he could even suggest that. They were standing at the top of the gully, at the point right before they would lose their protection from the wind – the point of no return. Jon, following the inner voice of his soul, as he calls it, emanating from his gut, said confidently, 'That's it, I'm turning around and heading back down.'

After a moment's thought, two others decided to go with him, while the other two said, 'No, we're going to do this thing, we're going for it.'

A couple of hours later, as Jon and the two others were carefully making their way back down the mountain, the two who had continued on up came hurtling down the ice and snow, barely missing Jon and one of the others. 'They almost took us out,' says Jon sullenly. The men fell to their deaths. Jon goes quiet for a moment and then says, 'I had thought they would just get frostbite, but no, they fell.'

In this and other life-or-death situations Jon has faced, the idea to stop and turn back didn't just appear in his head from nowhere, it always began with a feeling. A sinking sensation in the gut or belly, heaviness or dis-ease, sometimes nausea, but always with a feeling, a sensation, not an intellectual idea or concept.

Remember that intuition is *the productive use of unconscious information to improve decisions or actions*. But how do your decisions get access to your unconscious?

In Jon's case, his brain had processed a wealth of information about the wind, the temperature, how the team was feeling, and countless other things – all this valuable information was being processed unconsciously, without Jon knowing it. His brain had learnt over hundreds of previous climbs that particular patterns of data tend to result in particular outcomes; it had previously associated things that his senses were now processing with negative outcomes. When he saw how the wind was blowing off the ridge, and perhaps the way the snow had formed on it, and how the clouds were moving, and many other micro-patterns on the mountain, it triggered negative associations. These associations triggered negative feelings, those gnawing, sinking feelings in his gut, which made their way up to his decision-making.

Crucially, it matters *how* you access such information. We know of two primary ways to access and use unconscious information. One is to feel it, the classic gut-feeling definition of intuition, as Jon experienced on the mountain.

The other is to let your body directly take action, a physical action, such as running left, running right, passing the ball, or pulling up on the throttle of a Skyhawk jet fighter. Your body's movement has direct access to some of your unconscious.

People who've made intuition-based decisions often report that they 'felt it' in the gut, chest, or all over their body. The important point here is that they *felt* something that swayed the decision.

But wait: how can we feel the effects of unconscious information in our bodies?

You've probably heard intuition sometimes being referred to as a sixth sense, but there are actually eight sensory systems. Along with the five we're all familiar with – visual (sight), auditory (sound), olfactory (smell), tactile (touch), and gustatory (taste) – there's also proprioception (the sense of body awareness, knowing where your body parts are), vestibular (the sense of balance and motion), and interoception (our perception of all things internal to our body).

Interoception, which is often referred to as the eighth sensory system, tells us if we are hungry, stressed, sick, thirsty, need to go to the bathroom, or any other type of internal perception. Interoception provides a moment-by-moment mapping of the body's internal landscape across conscious and unconscious levels. But it also does a lot more than just tell you when you should eat, drink or go to the bathroom.

After a breakup or the loss of a loved one, people tend to say they are broken-hearted. Following a shock, they often say their heart skipped a beat. A case of nerves is frequently described as having butterflies in the stomach. Language often pairs body sensations and emotions like this. And there's a long history of talking about emotions in terms of the body, literally experiencing emotions *in the body*. This is called embodiment, or embodied emotions. As we feel an emotion, or even before we feel it, a whole array of bodily changes begins: the heart speeds up, blood pressure increases, we start sweating, our breathing changes, muscles tense up or relax.

The precise order of these events – whether the bodily changes come first or the brain-based emotions do – is a controversial topic. But the data does indeed suggest that bodily sensations, or interoception, are linked to how we feel emotions.

The distinction between emotions and feelings is not something that is universally agreed upon, and the terms can be used in different ways in different contexts. For the context of this book, I will note that emotions are made up of multiple things; these include not just our experiences, but also our behaviours and physiological responses. These elements coalesce, forming a vivid, multi-layered reaction to the situations we encounter in the world.

Feelings, conversely, can be considered as the conscious experience of emotions. Yet they are not confined solely to emotions. They extend beyond, allowing us to comprehend sensations that reside outside the standard emotional landscape – the gnawing feeling of the winter chill, the heavy feeling of fatigue after a long day.

In his 1994 book *Descartes' Error* neuroscientist Antonio Damasio describes a patient known as Elliot, who, after surgery to remove a brain tumour, displayed very strange and intriguing changes in behaviour. He seemed to be emotionally flat, and didn't display the normal range of emotional responses that he had had prior to the surgery. When it came to making decisions, he would endlessly debate the rational and logical pros and cons of each choice. Damasio describes asking him, 'What restaurant do you want to go to tonight?' Elliot answered, 'Well we could go to this restaurant, but I've heard it's been rather empty, which suggests it might not be good, but on the other hand if it's empty we will definitely be able to get a table, so we should go there.' The pros and cons would go on and on and on like this. Without the typical emotional response to the different options, Elliot had real trouble deciding what he wanted to do.

From case studies like this, and other experiments, Damasio proposed a really interesting idea that he called the Somatic Marker Hypothesis. The word somatic means relating to the body, so Damasio is referring to a body-based marker. His idea is that we mark options for a decision with emotions that are based in the body, or soma. Without emotions to help mark options as good or bad, we would all have great trouble deciding between options that are similar. Like Elliot, we would all get stuck in an endless loop comparing the pros and cons.

Damasio's idea of somatic markers that can aid decision-making is very similar to intuition. However, intuition crucially involves using unconscious information in the brain via bodily interoceptive sensations. In addition, the unconscious information is based on many instances of learning from the environment through repeated exposure and outcomes. Intuition is therefore not simply marking a response with an emotion, it is tapping into years of unconscious learning.

The Power of Intuition

In the fascinating world where science and emotions collide, things start to get pretty interesting. A myriad of experiments, including those conducted in my own lab, have unveiled the compelling truth that our bodies continue to respond to emotional stimuli even when we scientists manipulate them outside of conscious awareness. The art of rendering an image unconscious can be achieved through various techniques, and I'm not just talking about simply withholding an image from view.

We can present an image to an individual while using one of the many neuroscientific methods to render it unconscious, and their brain will still process it unconsciously. As we delve deeper in the coming chapters, the reality of unconscious perception will become more tangible, illustrating the mastery we've gained in the science laboratory. We can show you a fear-inducing image, such as a poisonous spider, and then stealthily suppress it from your conscious awareness.

And here's where things escalate: although you've never consciously perceived the spider, because we made it unconscious, we can examine your brain's activity to observe how its emotional regions react without consciousness. We can put a couple of tiny electrodes on your fingers and track minute changes in your sweat. When your brain is processing the deadly spider, you sweat more. This underscores the fact that our bodies are finely attuned to emotional stimuli even when we remain blissfully oblivious to their existence. In the next chapter you will see first-hand how we do this in the lab to create and measure intuition.

Interoception, therefore, is one of the main ways you can access unconscious information; your body will respond even if you don't know what it's responding to. Your internal organs and body parts are responding to unconscious signals, and interoception picks up on these signals.

In the collaboration between intuition and conscious deliberation, the unconscious often leads the way. Picture yourself settling in at a restaurant, only to be met with a nagging feeling that something is amiss. Heeding this sensation might spare you some unpleasant consequences, whereas dismissing it – as we've all likely done at some point – could result in unpleasant outcomes, such as an upset stomach or worse.

In this example, our minds aren't consciously weighing the pros and cons. Instead, much like Jon's experience atop the mountain, our brains rapidly and unconsciously process many cues from our environment: the aroma, the unkempt tablecloth, the ambience, the temperature, the demeanour of the staff, and countless other subtle factors. In mere seconds, this whirlwind of information is assimilated, associations are triggered, and our interoceptive system springs into action, turning these signals into a gut feeling.

This remarkable feat, this subtle-yet-powerful force, is a testament to the extraordinary capabilities of the human brain, and is what we commonly refer to as intuition.

It's important to note that this link between interoception, emotion and intuition has nothing to do with the gut's role in cognition. The gut microbiome produces neurotransmitters that *do* affect our mental health and cognition, but this is not what scientists mean when we refer to a gut response. We feel intuition in the body because of interoception, and not because of the microbiome. The gut-neuro connection is a whole other interesting topic itself, but won't be covered in this book.

Daredevil: how the blind can see

A man in his sixties, whom I'll call Tom, walks confidently but not quickly down a hallway. He is casually dressed, in a short-sleeved button-up shirt, and is followed closely by a man in a more formal-looking long-sleeved shirt and dark trousers. The hallway looks dated, like something you'd see in an unrenovated old university or hospital; it has a small sink and a fire extinguisher.

Placed purposefully along the hallway are two garbage bins, a camera tripod, a box of photocopying paper, a desktop inbox and a cardboard box, all sitting there like some indoor obstacle course. Each object has been placed in the middle or just off to one side of the hallway, so there is no way to walk directly down the hallway in a straight line. The more formal-looking gentleman following Tom is carefully watching his every move.

As the pair approach the first obstacle, a garbage bin, Tom rotates his hips, aligns his feet – one in front of the other – and steps to the side, up against the wall. He effort-lessly walks past the first garbage bin, and then past the second. Then, just as he looks as if he will walk directly into the camera tripod, he tentatively rotates the other way, stepping to his right and walking around it. Carefully, he then steps around the photocopying paper, and then, once again twisting his body back to his left, he sidesteps the cardboard box, straightens out, and finishes walking down the rest of the hallway.

To anyone watching this, it would seem completely normal, nothing out of the ordinary. But Tom is completely blind. He is considered clinically blind, failing all the standard vision tests. Scans of his brain have shown that due to two strokes, one not long after the other, his visual cortex – the part of the brain under the pointy bit at the back of your skull – was entirely destroyed. He typically walks with a cane, as do many other blind people, and requires guidance from sighted people. The man walking down the hallway behind him was actually running a research experiment, testing the ability of blind people to navigate. He was also there to help Tom if necessary and prevent him from falling.

This successful navigation through the hallway obstacle course is astounding. How can a blind man do this without tripping on the obstacles? This is an example of something called blindsight: a condition in which someone can respond to visual events and objects without being aware of them.

Tom's blindness is not due to damage to his eyes by the strokes. His eyes are perfectly fine. Tom's is a case of cortical blindness; he is blind because of brain damage. Here's the fascinating part: somehow, other parts of his brain that were not damaged by the strokes are processing the information from his eyes, unconsciously. Tom is not using sounds to navigate, or echolocation, or any other known method beyond visual input into his eyes. The information about the obstacles on the floor is there in his brain, he is just not aware of it. The striking point is that, despite this information being unconscious, he is still able to use it to aid his navigation when walking. This is a powerful example of unconscious information in the brain making its way into action, affecting behaviour.

In other tests, Tom was able to discriminate between still photos of happy and angry faces at rates well above guessing, but he could not tell the difference between faces with only neutral expressions; nor could he differentiate between pictures of black and white squares. The scientists performing this research demonstrated that Tom's amygdala (the tiny area of tissue deep inside the brain that many like to call the lizard brain because it automatically responds to things like emotion) was active in response to him being shown the emotional faces, even though he couldn't see them. The unconscious information was finding its way to the emotional parts of his brain from his eyes, without going through the usual routes in the visual cortex, which produces visual consciousness. Hence it was unconscious.

He couldn't see anything, but he could feel and act on the unconscious information in his brain.

There have been other cases of blindsight over the years and they provide some of the strongest examples of how we can use information in our brains without being conscious of it. Unconscious information in our brains is not locked away in some brain-dungeon with the key thrown out. Unconscious information leaks out into our conscious behaviour, our feelings, our choices. These examples of blindsight help inform our understanding of intuition. Intuition is simply utilising the way the brain is wired to let unconscious information leak through into feelings and actions. Using this information can give you an advantage over someone who ignores it.

In a very real sense, Tom was using intuition to navigate down the hallway. He had the unconscious information in his brain, he had learnt what it means, and he could access it to help him navigate through the world. He could access it in his decisions about the happy or angry faces.

Cases of blindsight, while tragic, show the power of intuition. The trick is to know when and for what you can trust your intuition, then practise using it.

Posting letters in invisible letterboxes

Imagine that when looking at your cup of coffee you don't see a cup, but a jumble of colours, vertical lines and edges, curves, and some textures. All these different visual elements are scrambled. They don't line up, so you can't see the actual cup. You look across the table to an apple, to what you think might be an apple at least, because as with the cup, you can't see its shape; you again see scrambled lines and colours, like one of Picasso's cubist paintings. It's as if a young child put your vision of the apple down on a piece of paper, cut it up with scissors, shuffled those pieces around, and glued them back together to make a collage. All the ingredients of the apple are there, they are just not in the right place. This is what people with Visual Form Agnosia experience when they look at objects.

Visual Form Agnosia is caused by damage to a very specific part of the brain, the part that's responsible for gluing together the different elements of vision into the coherent objects most of us see every day.

A young woman I'll call Jasmine presents a fascinating example of Visual Form Agnosia, which was brought on by an accident. Jasmine was renovating her home with her husband at the time, and if you've ever done a home reno yourself you'll know the chaos it creates. Up early one morning, she was having a shower in the mid-renovation bathroom. The window was closed, the ventilation fan was only half installed, and the room quickly filled with steam. Behind the shower curtain, hidden by the steam, was an older model gas water heater still in use, the type that sits inside the bathroom and not on a wall outside the house.

Jasmine, enjoying the hot water, was blissfully unaware that the room was filling up not only with steam, but also with carbon monoxide. As the old water heater burnt the gas to heat the water, it gave off carbon monoxide. Carbon monoxide is a clear gas that has no smell or taste. You don't know if you're breathing it in. Jasmine didn't notice anything amiss. But the gas is poisonous.

Carbon monoxide poisoning often starts with a feeling like having the flu. Jasmine began to feel a little dizzy, then things started to spin, much like with vertigo. Her heart began to race and she found it hard to breathe. Her chest muscles locked up and then – nothing.

Jasmine woke up much later in hospital, after passing out in the bathroom and going into a coma. She was lucky not to die. When she came out of the coma she felt okay – her vital signs were good. However, she had trouble with her vision, and couldn't see things clearly.

When she was up to it, the doctors began testing her for any neurological damage caused by the carbon monoxide poisoning. It very quickly became apparent that Jasmine had severe vision problems. She was then diagnosed with Visual Form Agnosia. She was seeing objects as scrambled cubist collages. Further tests and brain scans showed that Jasmine had suffered damage to the lateral occipital areas of her brain, on both sides. These are the parts of the brain just around to the side from the pointy bit at the back of your head, and they are known to be important for the perception of objects. They are the areas that glue the different elements of vision together, and Jasmine had lost a lot of brain tissue there.

One good thing about these types of brain injuries is that they are becoming less common each year. With advances in technology leading to safer gas heaters and better safety systems, there are fewer and fewer new cases of Visual Form Agnosia and other forms of selective impairment being caused by brain damage. This means that those who have this condition and are willing to participate in psychological experiments – as Jasmine is – are rare. She is flown around the world and put up in hotels to participate in experiments at different research labs.

During one testing session at a university research lab in Canada, a cognitive neuroscientist held a pencil in front of Jasmine and slowly rotated it to different orientations and instructed her to draw the orientation of the pencil. Even though the pencil was always held either vertically or horizontally, Jasmine's drawings looked like a spiky star, with lines pointing in all directions. She could not tell what the orientation of the pencil was, she was just guessing.

As the experiment went on, Jasmine became curious about the pencil. 'Let me see that,' she said, reaching out with her right hand to grab the pencil from the researcher. As she did this, her hand rotated to the precise orientation of the pencil and she easily grasped it.

The researchers were shocked. How could she grab the pencil so easily when all she could see was a scrambled collage? The researchers took the pencil back, tilted it to a different orientation, and asked her to reach out and grab it again. As she lifted her hand they were surprised to see it rotate to the exact orientation of the pencil once more. Again she grabbed it easily, tilting her hand at the wrist perfectly, opening her fingers to the precise width of the pencil.

Remember that when she was asked to draw the orientation of the pencil, she could only guess at it. But somehow, when Jasmine reached for it – when she made an action with her body – she, or her arm, seemed to access information about the orientation of the pencil. Her arm, it appeared, was not using her scrambled vision to grab the pencil, it was using something else.

In a more formal follow-up experiment the researchers, rather than holding up a pencil at arm's length, used a letterbox, a small item they had made themselves. It had a narrow rectangular opening not too different from that on a household letterbox, but it sat on a rotating contraption that the researchers could turn to any orientation they wanted. In the first step of the experiment, they asked Jasmine to draw the orientation of the letterbox opening, just as she had for the pencil. Each time she drew it, they rotated the box to a different orientation and asked her again. As with the pencil, the lines she drew were all over the place, all at incorrect, different orientations. She again seemed to be guessing.

In the second part of the experiment, the researchers gave Jasmine an envelope and asked her to post it in the letterbox. She picked it up without hesitation and posted it directly into the slot, without hitting the sides. The researchers then rotated the box to a different orientation, and once again she posted the letter without issue. How was she posting these letters when she couldn't tell or draw the orientation of the box?

This experiment is a striking example of how our visual and motor systems can use very different information. The information about the orientation of the pencil and letterbox was there in Jasmine's brain, but it was unconscious. She was simply not aware of it. All she saw was scrambled colour and lines out of place. But while she could not get conscious access to the orientation of the letterbox if she tried to draw it or talk about it, when taking the action of reaching out and posting the letter, her arm and hand could get access. Unconscious information was automatically making its way into her actions and guiding her behaviour.

I call this ability blindaction, a term to sit alongside blindsight. It is another remarkable example of the brain's ability to use unconscious information. Blindaction cases like Jasmine's are another demonstration of intuition in action. Think about unconscious information making its way into your actions next time you play sport or engage in other physical activities. Could there be information about the location of the ball, for example, in your brain that you're not aware of? And as you load up to kick it, could your leg and foot (really, your brain) tap into this information to better guide your kick?

Blindsight and blindaction are two instances of how our brain's unconscious information can affect our decisions and behaviour. When asked directly about this information, we can't access it. Tom saw nothing at all in the hallway, Jasmine saw the pencil and letterbox only as a scrambled mess of colour and lines. Jon didn't consciously know the mountain was too dangerous that morning, he just felt something in his gut. Somehow the unconscious information bleeds through so we can pick up on it with a feeling or act on it with an action. These dynamics form the basis of intuition. Intuition can be learnt, so that you too can do what Jason, Jon, Tom and Jasmine did: tap into the unconscious information in the brain and use it to make better decisions and perform better actions.

Blindsight and blindaction offer illuminating glimpses into the nature of intuition and its potential when honed through practice. We can sense unconscious information within our bodies, be it in the gut, chest, or the tips of our fingers – the faculty known as interoception. Those nagging sensations, the inexplicable uneasiness upon meeting someone, or the sinking feeling in your gut at a certain prospect, possess the power to convey valuable insights about events, as long as the five rules of this book (SMILE) have been satisfied.

This fusion of unconscious information and conscious decisions is intuition honed by experience, and it can guide us towards better decision-making and a deeper understanding of the hidden forces that shape our lives. Embracing this duality, we may find ourselves tapping into a wellspring of wisdom that lies just beneath the surface of conscious awareness.

Misintuition

The software team had been working on the presentation around the clock for three weeks now. They had made beautiful mock-ups, page after glossy page, perfectly prepared. Extra-large printouts rested on tripods around the boardroom. People were clearly nervous – fingers were fidgeting, bums adjusted in seats, glasses of water sipped to ease dry throats. Mike Evangelist, one of the software team, got up and rechecked all the mock-ups one last time, the screen-shots, menu options, and the piles of documents, all in order, laid out and ready to go. Yes: everything was ready, there wasn't anything else they needed to do.

The glass door to the boardroom swung open and Steve Jobs walked confidently in. He looked around the room momentarily, absorbing everyone and everything in an instant. Then with a calm precision to his movements, he walked over to an empty whiteboard and picked up a marker. He didn't look at any of the mock-ups on show around the room, or ask any questions about prototypes. Without explanation, without checking with anyone else, he drew a large rectangle on the board.

'Here's the new application,' he said. 'It's got one window. You drag your video into the window. Then you click the button that says *burn*. That's it. That's what we're going to make.'

Mike Evangelist and the rest of the software team were dumbfounded. This wasn't how product decisions were made anywhere else in the industry. How could product design simply follow one man's idea, one man's feeling about what would work? What if his intuition was wrong?

Product design typically involves group input for a

first-pass prototype. This is then followed up with either a focus group or user-focused testing and input. The groups discuss the pros and cons of different design choices and make suggestions, then perhaps more user testing takes place. But not here, not with Steve Jobs at Apple. Jobs is famous for saying, 'Some people say, "Give the customers what they want." But that's not my approach. Our job is to figure out what they're going to want before they do.' It brings to mind the classic quote from Henry Ford, of Ford Motor cars: 'If I'd asked customers what they wanted, they would have told me: a faster horse!'

Mike Evangelist was brought into Apple to help design Apple's DVD-burning program, which was eventually called iDVD, and it did follow Jobs' idea. This is an example of Jobs' supreme confidence in his own choices, his intuition about what would and wouldn't work. His obsession with intuition is on display in a quote from Walter Isaacson's biography of Jobs:

People know how to deal with a desktop intuitively. If you walk into an office, there are papers on the desk. The one on the top is the most important. People know how to switch priority. Part of the reason we model our computers on metaphors like the desktop is that we can leverage this experience people already have.

Jobs not only followed his own intuition, but also understood how customers might use it.

There are countless other examples of Jobs' fixation on making intuitive decisions, and wanting others to be able to use Apple products intuitively. How could he have known which things would work? The answer is: because he had mastery in his field, having spent years working on turning novel ideas into the reality of products. As we will explore more later, mastery in a given domain is one of the five rules for using intuition.

Jobs has been famously quoted by Isaacson as saying, 'Intuition is a very powerful thing ... more powerful than intellect.' As a young man, Jobs spent seven months in India in search of spiritual enlightenment. His time there taught him to use intuition, to trust it, to rely on it.

Tragically Steve Jobs passed away in 2011 of pancreatic cancer. His health choices prior to this had not been straightforward. His cancer was discovered by chance in 2003 from a CT scan for kidney stones that showed a shadow on his pancreas. This turned out to be a neuroendocrine islet tumour, a rare type of tumour that is slow growing and can typically be cured.

Strangely Jobs refused surgery for nine months after his diagnosis, wanting to first try non-invasive methods of treatment, including diet and lifestyle. According to Isaacson, Jobs said, 'I didn't want my body to be opened ... I didn't want to be violated in that way.' His resistance to surgery was apparently incomprehensible to his wife and close friends, who continually urged him to have the surgery.

Jobs' intuition had always come through for him at Apple. He had followed it in product design, and in the general direction for Apple – how could it not work for him for other things too? Why didn't the same reliance on intuition, after working so spectacularly at Apple, not also work for his health?

After Jobs' death, the *60 Minutes* interviewer Steve Kroft asked Isaacson, in respect to Jobs' decisions about his medical treatment, 'How could such a smart man do such a stupid thing?'

The answer is that while Jobs had mastered product design, development and innovation to a world-class level, allowing his intuition to flourish at work, he did not have the same mastery of health-related matters. Mastery, the second of the five rules for developing intuition, is crucial to practising intuition. Intuition is a learnt skill, and your brain needs to build the links between choices and outcomes. In addition, the context or environment in which you develop your intuition also matters, because learning is context-dependent. This means that intuition developed at work won't transfer well to other situations and locations. Environment, you'll recall, is the last of the five rules in SMILE.

Intuition, in other words, is not simply a matter of someone either having it or not having it. It is not so black and white. As we will learn in this book, intuition is complex, once we understand what it is and what it isn't. But fortunately, the rules for using it are straightforward and easy to follow.

I call the attempt to use intuition that leads you astray, as it did Jobs, misintuition. A misfiring of intuition. I feel that we need a word to distinguish productive, useful intuition, that follows the science, from the potentially more dangerous times when people think they're following their intuition but actually are not. Clear language helps avoid confusion about what intuition really is.

Jobs believed he was using his intuition when it came to his health, and so he blindly followed it. In fact, what he was following was a misguided feeling – his misintuition.

Understanding what intuition is, and when it will function and when it won't, is at the heart of my mission for this book. Yes, we can use and trust our intuition, but not always, and not for every topic.

When we follow our intuition without satisfying the five essential rules, we increase the likelihood of making suboptimal choices, taking ill-advised actions. In such situations, we're not guided by true intuition, but instead fall prey to misintuition.

Our brains love to trick us – they're good at it. They use hundreds of sneaky cognitive biases to crash our decision-making party. We'll discuss these biases in more detail later, but you've likely heard of their use by everyone from supermarkets to insurance companies to pull the wool over your eyes and make you spend more.

These cognitive biases are often confused with intuition, but it's simply not the case that every time we're nudged towards a decision without our full conscious awareness we're using our intuition. If we start down this path, then we'd be calling almost everything intuition; it's way too general a definition. It's important to be very clear about this if we want to make any real progress in understanding and developing intuition.

Cravings and addiction, however natural and necessary they may feel, are further examples of misintuition. And as we shall see, they can be particularly insidious. When the five rules of intuition are not met, we open the door to a range of different types of misintuition masquerading as intuition.

Measuring Intuition

How it all started

I have been studying how the brain processes unconscious information for twenty-five years. The aim in studies of consciousness is to get information into a person's brain unconsciously and then compare that to when they are conscious of it. I can then look at any differences in brain activity or behaviour, with and without consciousness, thereby figuring out the effects of being conscious of something.

My laboratory has chronicled a series of discoveries, shining a light on the enigmatic process by which the unconscious mind or brain processes information. My lab, and research from many others, has shown that a lot of information can be processed without consciousness. Through all these findings, scientists have illustrated how the subtle workings of the unconscious contribute to the decisions we make.

Before focusing on intuition, my work concerned the low-level sensory processing of elements such as colour or motion. But intuition, as we have heard, is a visceral experience – a sensation that involves feelings in the body, emotions, and gut feelings.

In 2013, I began working with a new student named Galang Lufityanto who had moved to Sydney from Indonesia to do a PhD with me. Galang really wanted to study intuition. I had some reservations, having never seen a good way to measure it scientifically. At that stage, the science and theories of intuition seemed mixed at best, but mostly there weren't any. Case studies like Tom's and Jasmine's, which we saw in the previous chapter, were fascinating, as were the many anecdotal stories like Jon's that I kept hearing, but what was needed was a way to study intuition in anyone, at any time.

Galang and I began by talking about how people had defined intuition up to that point, and about what intuition was and what it wasn't. We discussed the lack of empirical research and why there were so many non-scientific books on the subject, despite no clear definition. Many non-scientists defined it as something beyond science: a sixth sense, something akin to magic.

The lack of good metrics and science notwithstanding, there seemed to be a huge appetite for information on intuition, in people from all walks of life – business, sport, spiritual, and even the military. Why, then, were so few scientists looking into it? Scientists were still arguing about how best to define it, whether it was a positive or negative phenomenon, useful or misleading. Some even doubted its existence. Hardly any were trying to develop ways to measure it in the lab.

Many scientists believe that you need a clear definition of something before you can launch into empirically researching it; that without an agreed-on definition, you'll end up trying to measure different things, thereby adding confusion to a developing field and preventing breakthrough discoveries. But here's the twist: often, more data is needed before people can agree on a definition. It's a chicken-and-egg impasse: which comes first, the definition or the study to obtain data?

One way out of such an impasse is to use a working definition of something: a temporary description that's enough to get the research going, to inspire or frustrate people enough to formulate their own theories and models and start running experiments. So, to kick things off, my working definition of intuition was 'the learnt productive use of unconscious information for better decisions and actions', very close to the definition I use in this book.

The next step for Galang and me was to think about how we could test that definition, by actually measuring it. I'm a huge fan of Aristotle's 'first principles' thinking. A first principle is a basic assumption or element that cannot be deduced from or further broken down, like the basic units of matter. First principles thinking is at the heart of science, and is also crucial to good engineering. Going back to first principles can often reveal problems in how things are currently being done; for example, in industries using old legacy methods of manufacturing that have been handed down through generations. By starting afresh and combining the essential elements or principles of the process, new, more efficient, cheaper or faster methods can be devised.

You may have heard Elon Musk talk about his first principles thinking for designing and building rockets at SpaceX. This involves going back to the basic building materials of rockets, looking at the cost and availability of steel, rubber, glass, and all the other materials that go into making a rocket, and then rethinking the way things had been previously done. SpaceX is now one of the companies at the forefront of space exploration, and given Musk's success, first principles thinking has again become a popular way to get to the fundamentals of something. Rather than buying a pre-made rocket, you go back to the drawing board and see what it will cost to build it from the ground up. Not only can this be cheaper, but often and more importantly, you will not inherit the bad habits, outdated tech, or mistakes from the old designs.

This is the same for psychology and neuroscience.

First principles thinking can be applied to almost anything. Galang and I used it for measuring intuition. We had spent months whittling down a working definition of it, outlining all the basic ingredients of what intuition is and what it is not. As part of this thinking process, we intentionally ignored the existing methods others had claimed might measure intuition. The idea was to start with a clean slate, to avoid any bias in how the study of intuition had been approached so far.

It was apparent that one of the key ingredients was decisions that are based on unconscious information: knowing what without knowing why. A second was the speed of decisions, like a quick gut response. The next step was to review the neuroscience and psychology literature to see what tools and methods existed to measure each of these ingredients of intuition.

Emotional inception

In the 2010 film *Inception*, Leonardo DiCaprio's character, Dom Cobb, specialises in a very precise type of security: subconscious security. He is charged with the task of implanting, or incepting, an idea into the unconscious mind of a company CEO, so as to influence his decisions. To do this, Cobb and his team hack into dreams.

When we started working on intuition in the lab, this movie played more and more into my thinking. Because one of the most critical elements in our definition of intuition was that the knowledge being tapped into is unconscious, we needed to find a way to supply the unconscious information and get it into the brain, then see how people could use it in decision-making. If we could manage this, we would have the first building block for creating intuition in the laboratory. We needed an easier way to incept information into someone's mind than hacking into their dreams. So here was the seemingly daunting task: how could we do what Cobb's team did without having to go to the extremes of his methods?

As I mentioned, my lab and I had studied consciousness for more than ten years, so we were familiar with the different methodologies for controlling it. Consciousness research is a little like a real-life version of *Inception*, minus the car chases and guns. There are a few different ways of incepting things into the mind that do not require dream-hacking technology.

We chose a method that allows control of someone's consciousness of an object presented to one eye.

When you look around the room, presuming you have sight in both eyes, you see in beautiful, three-dimensional depth. There are objects that are closer to you than others and you perceive this immediately and effortlessly. This is what allows you to catch a ball or to reach out and grab your coffee.

The way your brain does this is by fusing together the vision from each of your eyes. Because your eyes are separated by your nose, each eye gets a slightly different view of the world. You can easily demonstrate this by holding up a single finger about a hand's distance from your face. Rather than focusing directly on your finger, relax your eyes and look at the wall, out the window, or whatever is across the room from you. You should now see two fingers in front of you. They will look a little strange and transparent, a separate image from each eye, but that's normal. When you do this, you're catching a glimpse of the two different data streams your brain is receiving from your eyes. The brain takes these two different views and stitches them into the three-dimensional world you see as you look around.

However, if two very different images are artificially presented, one to each eye (something which rarely happens in normal life), the brain cannot fuse them together. Your visual system is thrown into a state of rivalry. Neuroscientists call this binocular rivalry.

Your brain has a very curious solution to this rivalry. Suppose the two objects are a banana and a green apple, each quite different from the other in shape, texture and colour. When your brain is confronted with these under controlled lab settings, you will first see one image, then all of a sudden that image will disappear and you will become aware of the other. The two objects will battle it out in your brain to win your consciousness. When the experiment is set up in the right way and the two images are of equal intensity, these amazing vacillations of consciousness will continue indefinitely.

If, however, you make one of the objects – let's say the banana – very bright, colourful and flickering, then it will dominate and win the battle for your consciousness. The beautiful vacillations of consciousness stop and you will just see the amped-up, flickering super-banana. While you see the banana, the apple is pushed into your unconscious, and you might never see it. But the image of the apple is still hitting your eye, it is still being processed by your eye, and that information is still being relayed down the optic nerve, first to subcortical processing, and then to the cortex.

This method of halting binocular rivalry oscillations is called Continuous Flash Suppression. The flashes of brightly coloured objects continually suppress whatever is presented to the other eye.

Continuous Flash Suppression is the perfect tool to create inception without needing to hack anyone's sleep. This method is well-suited to intuition because we can present information to one eye, such as the apple, and then render it unconscious, in a similar vein to blindsight or blindaction.

Another important ingredient of our definition of intuition was that it is *felt*. Generally people report that they feel it in their bodies, just as Jon Muir did. And you'll recall that Tom's amygdala was active in response to emotional images, despite him being completely unconscious of them. He could tap into this unconscious emotion and use it. People feel intuition without knowing it or being conscious of it, therefore, we chose to use emotional information rather than just perceptual information, like a banana, for inception. I call this Emotional Inception. All we had to do to achieve Emotional Inception was to swap the apple for an emotional image, such as a deadly spider – something most people would have an emotional response to. So now we had an easily controllable way to present emotional information (the spider), and then take away conscious awareness of it with the supercharged flickering banana: Emotional Inception.

We know from brain-imaging experiments and other types of physiological measurements that parts of your brain will process the spider despite you not seeing it. The deadly spider will have a brief effect on the amygdala. This small part of your brain is tucked in close to the centre, deep inside the wrinkled cerebral cortex shell that wraps around the outside. The amygdala is best known for processing and responding to fear, but it also processes other emotions. Right after the image of the spider goes into your unconscious, the amygdala will momentarily be more active. That activity will soon die down and things will go back to normal.

Now you may wonder how we know when we have successfully created Emotional Inception. First, we measure people's physiological skin conductance. When our amygdala processes something fearful – or for that matter whenever we are emotionally aroused – we sweat just a little bit more. I'm not talking sweat dripping onto the floor, like when you're doing a workout, but just a small, unnoticeable increase. This makes the skin more conductive to electricity. By running a very low, undetectable voltage over the skin, it is easy to measure this physiological change in sweat. During Emotional Inception, we see clear evidence of emotional arousal, even though the emotional images are never consciously perceived.

Scientifically, it is important to check that participants in the research really are unaware of these images, and to this end we always include a stage in which people must try to guess what the image of inception is, whether it's a puppy or spider, or a blue or red square, or whatever. Crucially, the data shows that people trying to report which image they have been shown perform no better than if they were guessing. Therefore, the data confirms that people don't consciously know what that image is.

In our mission to measure intuition, we now had a working definition and a method for supplying unconscious information. The next element we needed was a decision-making component – a way of testing decision-making that was fully conscious.

Scientists have studied decision-making in different ways and one of the more fruitful methods uses extremely simple decisions. I'm talking about decisions that are so simple they almost seem silly, such as identifying whether an object is moving to the left or to the right.

Such simple decisions are very powerful scientifically precisely because of their simplicity – they don't involve lots of variables that can interfere with the experiment. A more complex decision, for example, about which car to buy involves many variables, apart from obvious things like cost. Different people are going to like some cars and dislike others; maybe they once had an accident in a Toyota, or perhaps their childhood car was a Volvo. These factors bring extra emotion to the decision-making process, which can add noise to the experiment. They make it more difficult to understand how people make decisions.

To solve this problem, neuroscientists use clouds of little dots on a computer monitor. Each dot moves independently, and all a study participant has to do is decide which overall direction the cloud of dots is moving in. It's a bit like looking at white 'snow' on an old-school broken television, with little pixels moving all over the place. We can manipulate the pixels on the computer, making most of the dots move randomly so that only a small amount are moving identifiably left or right.

It takes a moment to figure it out, because if most of the dots are moving randomly they disguise those moving left or right. We decided to use this conservative and super-simple method to study decision-making and intuition.

We now had the fundamental ingredients to create and measure intuition in the lab. We had a definition. We had Emotional Inception. And now we had a conscious decision-making task that let us measure decision-making cleanly and simply. However, there was one other final element we needed to measure intuition in the lab.

Remember that intuition is the *learnt* productive use of unconscious information to improve decisions or actions – accordingly, we needed to have something for participants' brains to learn, even if this learning was unconscious. (It's helpful to note at this point that instincts and reflexes are things we are born with; we come into the world with such biological responses intact. We'll deep-dive into the differences between instincts and intuition, rule three of SMILE, later.)

In this design, we wanted the brain to learn the association between the emotion of the unconscious image and the conscious-decision outcome – just as Jon's brain had learnt the association between hundreds of variables on the mountain top and success or danger when climbing. To achieve this, we set things up in such a way that the emotion of the unconscious images was either positive – using images of puppies, flowers, and the like – or negative, using images of a deadly spider, snakes, or a shark's open mouth.

And here is the bit that the participant's brain had to learn: the type of image was always linked to the answer to the super-simple moving-dot decision (left versus right). In other words, we presented positive or negative images depending on the correct answer to the decision task. For example, if the answer to the simple decision was left on a given trial, then for that person we would *always* present a negative image to their unconscious when the dots were moving to the left. In this way, Emotional Inception was unconsciously giving the participant's brain the answer to the moving-dot task. Of course, this association had to be learnt, it wasn't innate.

We had a hundred participants in this experiment, and we found that when decisions were accompanied by positive or negative unconscious emotional images (inception), tied to the outcome of their decision-making, people learnt to make more accurate decisions. That is to say, the presence of these *unconscious images improved decision-making*. But this wasn't immediate. The learning took time.

The images helped people answer the simple left-versus-right decision task, in the same way that the emotion of the facial images helped Tom decide which image was in front of him.

Not only did people get more accurate in their decisions (a higher percentage of correct answers), but they also got faster at making decisions. We asked the study participants to report their level of confidence after each decision and found that they reported being more confident about their choices when they had been presented with the emotional images than when there was no relationship between the emotion in the images and the correct decision. When things were random like this, the relationship between the emotion and decision could not be learnt. In other words, when learning to tap into unconscious information and use it in decision-making, it wasn't just the presence of the emotional images that mattered, it was the learnt association.

After developing that method of measuring intuition, I spent the best part of the following decade refining a robust theory of intuition. What it is, what it isn't, when we can trust it, when we can't. The five rules of this book emerged naturally from the deep synthesis of not just the science from my lab, but the work of hundreds of others over the past century. Work that has shed light on emotions, anxiety, depression, the science of learning, addiction, impulses, the fascinating and often frustrating psychology of probabilities, and the importance of the environment in learning and decision-making. It's a bold and exciting endeavour to harness the power of science to fundamentally change how we make decisions for the better.

I strongly believe that defining intuition as the *learnt productive use of unconscious information for decision-making and action* is the most useful and practical definition of intuition we can have. It provides a clear path forward for scientific research. We can explain it right now with all we know from psychology and neuroscience. Definitions such as those claiming intuition is a magical sixth sense that taps into information in the ether are not helpful. They don't provide an avenue for scientific discoveries, or ways to put those discoveries to good use. Part of my aim in writing this book is to catalyse the science of intuition into a major discipline that can be applied to every decision everywhere, to improve our lives by improving our decision-making.

Measuring intuition in the lab:
the experiment

You walk hesitantly into the small dark room. The walls are all painted black and there's only a dim light coming from somewhere up in the ceiling. You take a seat at a small, almost-empty black table. On the table sits a computer monitor, about an arm's length from you. Right on the edge of the table sits a device not unlike the machine an optician uses to test your vision, with a custom-moulded chinrest. But unlike at the optician's, everything here is black.

The gothic optician's device has four little mirrors set at different angles, reflecting what little light is in the room back at you. The technical name for this is a mirror stereoscope. You very carefully lower yourself into the chair.

Once you're seated and stable, I say, 'Take a look through the mirrors,' and you lean forward and peer into the device. Two small mirrors line up perfectly with your eyes; you immediately see two white squares in the mirrors. The mirrors show you two sections of the computer monitor.

I adjust the mirrors for you, changing their angle slightly. The two squares you can see through the mirrors snap together, instantly becoming one clear, precise white square. 'Okay,' I say, 'we're ready to start the experiment.'

You take a deep breath and it begins.

Immediately you're almost overwhelmed by bright flashing colours. Scrambled collages of random shapes of all different colours flash rapidly, one after another, overlapping and replacing each other. This is the continuous flash suppression that enables Emotional Inception; it's continuous because it freezes those binocular rivalry alternations.

Unbeknownst to you, hidden behind the bright colours of the continuous flash suppression is an emotional image; on this particular trial it's a picture of a venomous spider, but on the next trial it might be a picture of a puppy. But remember that you never consciously see the image, it is suppressed outside of awareness, so you are blissfully unaware.

Being in a dark room with dark walls makes the bright colours seem all that much brighter, but their flashing is not uncomfortable. It's on for less than a second and then stops. The screen goes black, and once again you're engulfed in darkness.

At the same time, just to the right of these flashing colours, is a snowstorm of small moving dots. They're going in all directions like thousands of radioactive glowing ants crawling over a nest – up, down, left, right – and your job is to make a decision about the overall direction these dots are moving in. Not the direction of each individual dot, there are way too many for that, just the overall direction of the majority. It's difficult to say, because many of the dots are moving randomly, but somehow you decide that there are slightly more dots moving to the left. You hit the button marked with a left arrow on the keyboard to report your answer.

So far so good. A little strange perhaps, but straightforward.

You continue with another trial, another decision, then another and another. This isn't that hard, you think, and you let out a sigh as you start to relax. You're feeling more confident that you can do the experiment. You wriggle a little to get comfortable in the chair as you go on to finish it.

This is what it's like to have your intuition measured in my lab. It might seem overly reductionist, but this represents the first lab-based demonstration that unconscious emotional information can be used for decision-making. This experiment shows that we can tap into unconscious information in our brains, associate it with conscious information and use it to make better decisions. In other words, that intuition is real. It can be measured scientifically.

PART TWO

The Five Rules
for Intuition

Self-awareness

Smile

Feeling Emotional? Don't Trust Your Intuition

Confusing love and intuition

My date and I had filled out all the paperwork, including insurance details, signed the documents, and picked up a harness. Things were still a little awkward – every comment, action and response was scrutinised, as they only ever are on a first date or in a job interview. We squeezed on our climbing shoes, dipped our sweating hands into the bags of powdery white climbing chalk, and were ready to hit the indoor climbing walls.

As first dates go, this was up there with the scariest. We were jumping straight into the deep end. Everything was on display: strength, bravery, and tight harnesses pulling in all kinds of places. We clipped in, myself as the climber and my date as the belayer, holding my life in her hands. Well, at least it felt that way. What an acceleration of trust, speeding straight through all the usual first-date small talk, the psychological games of poking and prodding each other, to this.

Neither of us were competitive climbers, but neither of us were novices either. I used to climb a lot when I was younger, but hadn't for years. With every metre higher I could feel the little butterflies of fear flapping their wings in my stomach.

Then the moment hit, everything slowed down and time moved to a standstill. I could feel the textured chalky, fake rock handholds moving under my fingertips, at first just by millimetres, then by centimetres. Then my fingers slid, my grip failed, and instantly I knew it was over.

Adrenaline, embarrassment and failure all go through my head before my fingers leave the handholds and I start falling. It's like watching a train wreck in slow motion. I'm off the wall, floating in space, before the elastic climbing rope does its job and absorbs my weight as the harness cuts into my thighs. I swing in and hit the wall in front of me with a thud.

My date lowers me to the ground, where I catch my breath and balance. My fingers are stinging and my arms are burning but I feel amazing, so alive, full of adrenaline. We're having a fantastic time. I look into her eyes, knowing she feels it too, and smile.

We swap places. Now I am belaying for her, dutifully pulling the slack rope through the belaying device, trying to be impressive. I'm ready for it, ready at any second to minimise her fall. And then *woosh*: she's off the wall, falling until the rope again does its thing and absorbs her fall. I hear her feet hit the climbing wall as she swings into it. I lower her to the ground, and through gasps of air she looks into my eyes knowingly and smiles.

As it turned out, we were not suited for each other. The electricity of that first date, the sensation that the air was crackling and dancing, proved elusive in our subsequent encounters. We could not rekindle the spark of that initial connection, and eventually accepted that our relationship

lay in friendship. She later revealed that her affections typically gravitated towards women rather than men, yet the
powerful chemistry we shared on that first date had ignited
a belief in her that things would work between us.

Those one-off strong feelings from that date stuck with
me for years. Why had things felt so right then, but not afterwards? I would often wonder how we got the chemistry so
wrong. I finally put two and two together and realised that
we had been the victims of something called misattribution
of arousal.

Humans are very bad at knowing what our emotions are
and where they come from. My date and I had confused the
adrenaline of climbing – the racing hearts, the sweaty palms,
the sheer excitement – with feelings of attraction. Our brains
couldn't tell where these feelings had come from: the climbing
or the chemistry between us. This is a phenomenon that
the producers of shows like *The Bachelor* and *The Bachelorette*
know all too well and use to manipulate participants.

In the now-famous study that's referred to as 'the rickety
bridge experiment', researchers at the University of British
Columbia had male participants walk across a high, unstable
suspension bridge. While on the bridge, they were met by a
female actor or confederate, who was part of the study, and

asked to write a short dramatic story about a picture she showed them. When she collected their story she gave them a phone number, a fake one, and told them to call her if they had any questions about the experiment.

Just nearby was another bridge, lower, wider and more stable, where the experiment was repeated. Those who took part in the experiment on the high, rickety bridge were not only more likely to call the female in the experiment, using the false phone number, they also used more sex-related language in the stories they wrote.

The researchers interpreted these results to mean that the males in the study had become confused about why they felt emotion or arousal. Those high up on the unstable bridge would have felt a little anxious, nervous, maybe some adrenaline, due to the situation, but they probably were not aware of why they felt these things. Just as my date and I had done, they misattributed these feelings to a response to meeting the girl on the bridge.

In other words, when we feel aroused, be that arousal positive or negative, we are not very good at knowing where it comes from. Other studies have been done in which males were asked to do physical exercise, such as running on the spot (nothing scary like the high rickety bridge or rock climbing), and then immediately afterwards, rate the levels of attractiveness in females in a short video. The men who had exercised rated these females as more attractive, compared to males who didn't do any exercise.

Even in these less extreme experiments, people confused the cause of their elevated heart rate, sweat and warmer body temperature. While these were simply due to physical exercise, participants had a tendency to attribute them to attraction towards another person.

When it comes to intuition, if our system is flooded with strong emotion, positive or negative, or with anxiety, we run the risk of being a victim of arousal misattribution. For the same reasons that you shouldn't rock-climb on a first date, you shouldn't use your intuition when you're full of adrenaline, depressed or anxious: you could easily misinterpret your feelings as coming from intuition and not from their actual cause.

So if you're feeling anxious about crossing a high bridge, that's not necessarily your intuition. If you've just done a workout or had too much coffee, your heart might be racing; that likewise isn't your intuition pinging your emotions, it's just your elevated physiology. Strong emotions, substances and exercise can put your body into a state that can easily be confused with intuition – don't misattribute these things to your intuition.

This is why the first rule in SMILE is so important. S is for self-awareness of emotion. *Use your self-awareness to check your emotion before practising intuition.* You should not use your intuition when you are emotional, anxious, depressed, or full of adrenaline. Strong emotions will not only over-power your sensitivity to the subtle signals of intuition, they might also result in misattribution, thereby leading to a bad choice – misintuition.

Noise drowns out intuition

Imagine you're at a party. It doesn't have to be a crazy raging party, just something with music and more than twenty people. You're standing there, leaning against the wall, drink in hand, enjoying the vibe and the music. Across the crowded room you spot an old friend you haven't seen in years. You shoot up your right arm to wave, but they don't see you. Someone else has their attention. You call their name loudly but the music and bustling groups of people through-out the room, talking and laughing, are just too loud.

You manoeuvre through the room, dodging people and furniture, to finally arrive beside your old friend. You say, 'Hi!', but the room is still too loud; all the talking heads and music drown out your voice. Mustering up a louder voice, you shout, 'Hey, long time no see.' Finally, they turn around, looking surprised but happy. You keep talking. 'What have you been up to?' They shrug, cupping a hand over an ear. You point to the kitchen, which looks empty.

Walking together into the kitchen you repeat yourself: 'What have you been up to?' Finally, you can hear each other, you can talk without the noise drowning out your voices.

It's the same with intuition. Any strong emotion you're feeling will drown out the subtle voices from your unconscious. Not only do you have to be wary of emotional misattribution, you also need to be aware that strong emotions will drown out your intuition, as loud noise drowns conversation.

Let's go back to Jon Muir on the mountaintop. As he and his team were ascending Everest, Jon's senses rapidly processed all the features of the mountain: the wind, the softness of the snow, the sunlight, the energy of the team, their body language, and hundreds of other details.

Jon's brain then introduced a feeling based on the associations between all those different things, and any positive or negative outcomes he had previously experienced in similar situations. Notably, he was sensitive enough to pick up on these associations, which can often be very weak. If he had been highly emotional, he may have missed his bodily sensations warning him about the dangers of the mountain.

Only use intuition when you are in a good clear state of mind. While strong positive emotion will likely block your sensitivity to the signals of intuition, medium-to-weak positive emotion – just being in a good mood – can be good for intuition.

Research into the effects of mood on intuition has used a measure known as the semantic coherence task. This involves participants looking at three words and rapidly deciding whether or not they have something in common. For example, the words *salt, deep* and *foam* are all associated with the sea. Participants are only shown sets of words that are linked by an overall concept, never unrelated words. The research shows that people sometimes answer that the words are linked even when they don't know what the linking concept is. This knowing that words are semantically linked without knowing how has been described as an intuitive judgement.

Scientists have also shown that putting people in a good mood improves their performance in the semantic coherence task, and that putting them in a bad mood makes their performance worse. For this study, participants are asked to first recall a recent event in their lives, either a positive or a negative one, and write down the emotional aspects of it. Doing this tends to bring that emotion back to life. They are then asked to do the semantic coherence task.

What we don't know is whether the effects of mood extend beyond tasks involving words and their meanings. We don't know, for instance, if such mood induction procedures would change performances in our measure of intuition using Emotional Inception. Or whether Tom would still be able to walk so effortlessly down that hallway when in a bad mood.

However, this doesn't change our first rule: check your emotional state before you turn to intuition. If it's too high in either direction, good or bad, turn to rational, conscious logic instead.

Anxiety and depression

Each person's experience of anxiety can be different but there are some commonalities. Tightly clenched fists, the sound of your heart pounding in your chest, extreme focus on the impossibility of what is happening. The insistence on thoughts such as, 'This can't happen, it can't continue like this.' The dread and the fear. The forever feeling, on repeat, that the world is immediately coming to an end, at this very moment, over and over again. These feelings of anxiety can be so strong you *have* to believe them. How can something so strong not be true?

At last the imploding chest and pounding heart settle and you feel tired, almost sleepy. The worst of it is over now, and there's a dim light again at the end of the tunnel. Things are more bearable, the world is no longer ending, and you can breathe once more, long slow breaths.

Anxiety can stick around as a general state and an ongoing feeling, or it can flare up like a panic attack. I have had a few bouts of anxiety in my life, and they really are some of the most horrible things I've experienced.

When anxiety is very strong and disruptive, it's not surprising that cognition undergoes radical changes. Short-term memory takes a massive hit; potentially threatening things in the environment grab and hold our attention, making it hard to concentrate on anything else. The early data suggests that intuition is impaired by anxiety, too.

Researchers looking into the effects of anxiety on intuition once again used the semantic coherence test. The scientists running this particular study first put people into an anxious state by getting them to read an anxiety-provoking text that depicted negative circumstances that were out of their control. Right after that, the participants were presented with scary images of snakes, spiders and sharks on a computer screen. Doing this over and over with different text and images increases anxiety ratings. The participants were then asked to perform a few trials of the semantic coherence task, before being shown more nasty images and then doing further semantic tests.

Ethics committees carefully vet studies of this type, and anyone with pre-existing psychological or neurological conditions does not usually participate. The induced state of anxiety is only temporary, and researchers check to make sure people are feeling okay again when they leave the experiment.

The participants in whom anxiety was induced

performed significantly worse at the semantic coherence task than those who were not put into an anxious state. The authors of the study interpreted this finding to mean that *being in an anxious state reduces your capacity for intuition.*

Why would anxiety have this effect? The authors of this study link it back to the idea of emotional misattribution. When we are anxious, we confuse the source of our feelings. The study participants confused their induced feelings of anxiety with any feelings that arose from the semantic coherence task, which they would otherwise have interpreted as a signal that a single concept linked the words in the task. I think a state of anxiety also creates emotional noise, effectively hiding the signals of intuition. For these two reasons, we have trouble detecting the subtle emotional signals of intuition when in an anxious state.

It's important to remember that this research has only been done with temporarily induced anxiety states. We don't yet know how people who have been clinically diagnosed with anxiety would perform on the semantic coherence task. But I predict that clinically anxious individuals would, for the same reasons outlined above, do poorly with intuition.

*

The other debilitating mental health condition that has been shown to have strong effects on decision-making is depression. People suffering major depression often cannot make up their minds. They tend to get stuck on negative thoughts, often overthinking the causes and consequences of their state, situation or mood. These loops of tunnel vision are called rumination, and they are more than just a symptom of depression – they are a mechanistic part of it: they maintain a depressive state, making it very hard to take any action to change things. Depression has been shown to impair problem-solving, decision-making and intuition.

To test how depression might disrupt intuition, researchers once again used the semantic coherence task. But for this study, rather than trying to temporarily induce depression in experimental subjects, they recruited people who were currently in a clinical state of depression. At the research laboratory, participants' depression was verified by an interview, and they were then given the semantic coherence task. The data showed that the depressed people performed significantly worse than people in the non-depressed control group. This result suggests that depression, much like anxiety, disrupts intuition; hence, there is evidence to suggest that both anxiety and depression disrupt intuition.

It is worth keeping in mind that all this research was done using the semantic coherence task – a test which, as the name implies, is specific to words and their meaning. A related study used, instead of words, coherent and incoherent pictures for the task, which was otherwise the same as the word version. But here the researchers found the opposite pattern – depressed people did a little better than the non-depressed. This makes sense, as we know that negative emotion can boost visual processing, making you more sensitive to seeing very faint objects in pictures.

The opposite outcomes of these two types of coherence tests are important, as the difference begs the question of whether the semantic coherence task really measures intuition or something largely based on language. We know that verbal rumination is synonymous with depression, and that rumination is word-based, so perhaps rumination could be disrupting performance by depressed people in the semantic coherence task. Unfortunately, no one has yet done the research to clear this up.

How to know if you are too emotional

Here's a scene from a video posted online. It's from the city of Cocoa, Florida, not far from the space-launch site at Cape Canaveral, and it shows a young man, Jamel, manoeuvring his way over a backyard fence. The fence is higher than his waist, but not so high that he needs to climb up and over it, making it that awkward, in-between height. In the grass on the other side of the fence Jamel stumbles and trips, disappearing behind the long grass for a minute. When he finally gets to his feet and starts to walk again he limps slightly. He goes down the hill towards a lake, eerily still with the reflection of the surrounding grass and trees. He pulls off his shirt and splashes into the water. The emanating ripples break the reflection as he begins to swim out towards the centre.

Next thing, Jamel's head emerges as a small dot in the middle of the lake and a young male voice is heard yelling, 'You fucking junkie, we're not going to help your arse!'

The footage is from a hand-held phone; the wobbles and jerky moves make that clear. We hear more commentary from those holding the phone: 'Ain't nobody going to help you, you dumb bitch.' The grass going down to the lake at the bottom of the video shows just how close the group of teenagers we can hear are to the water's edge. 'He keeps putting his head under ... wow.' Jamel's head is now disappearing and reappearing in the water. 'Why you scared to see a dead person?' a voice asks.

That's when you clearly hear Jamel shout something – a wail or cry for help. The small dot on the still lake surface, his head, disappears once more.

'Oh, he just died,' yells someone, and the group sneers, laughs and chuckles. 'Your buddy has been under there for a while now, he ain't coming back up now.'

'Damn, where he at?'

'We just saw somebody die, and we didn't even help him.'

'Yeah go help him, he dead. Buddy not coming back up for real.'

The video is as hard to watch as you might think. Night terrors still come for Gloria Dunn, Jamel's mother, who regularly wakes up gasping for air, feeling like she is drowning with her son.

Why didn't those teens help Jamel? Why didn't they phone for help, or at least call the police?

I've recounted this harrowing story to highlight the importance of emotional intelligence and emotional awareness. There has been much talk in recent years of a decline in emotional intelligence, but what is this exactly?

Emotional intelligence is a general term given to a collection of qualities or characteristics, from someone's own emotional awareness to the awareness of others and the ability to manage or harness emotions. High emotional intelligence predicts a huge range of positive things in life, from job performance, wellbeing and stress management to academic performance and social relationships. Low emotional intelligence is reflected in the lack of response to suffering seen in the tragic case of Jamel.

The general decline in emotional intelligence we've seen over the past decade or so has been linked to increased use of technology, particularly social media. Although it's hard to tease out the cause-and-effect relationships in the studies that have been done on this, most of the authors propose that the rise in young adults' use of social media might be responsible for the decline in emotional intelligence. These results also fit well with the many general findings showing a clear association between time spent on social sites and negative wellbeing or psychopathology.

Awareness of your feelings is not only a critical part of emotional intelligence, it's also critical to the safe and productive use of intuition. Emotional awareness is a component of your emotional intelligence, and it's important here because you need self-awareness to know if you're in an emotional state and whether it's safe to practise intuition.

What is emotional awareness, and how do you get it if you don't have it?

As I've said, emotional awareness is a part of emotional intelligence, which is typically measured with questionnaires. For example, the Trait Emotional Intelligence Questionnaire (short form), includes these specific statements to measure emotional awareness:

1. Expressing my emotions with words is not a problem for me.
2. Many times, I can't figure out what emotion I'm feeling.
3. I often find it difficult to show my affection to those close to me.
4. I often pause and think about my feelings.

Respondents give a number from 1 to 7, with 1 for 'completely disagree' and 7 for 'completely agree'. I've included them here just to give a feel for the questionnaire: answering them outside the context of all the other statements in it will not measure your emotional awareness. There are many online emotional intelligence tests you can run; however, be warned that they may not all be valid, so take your score with a grain of salt.

Emotional awareness is a complex trait that is shaped by a multitude of factors. While some people seem to be naturally more attuned to their emotions, others struggle with identifying and managing them. So, what are the key determinants of this awareness and can it be improved with practice?

One of the most important factors is early life experience. Children who grow up in environments that foster emotional expression and the development of emotional regulation skills are more likely to develop higher levels of emotional intelligence and awareness. If a child is encouraged to label their emotions and talk about how they're feeling, they will better understand their emotions and learn how to manage them.

Genes also play a role in shaping emotional temperament, which can impact emotional intelligence and awareness. Some individuals are predisposed to experiencing intense emotions, others are more even-keeled. That said, genes and childhood experiences are not the be-all and end-all of emotional intelligence, which can still be improved through targeted training and practice.

One area where individuals can focus their efforts to improve emotional intelligence is in developing the emotional awareness component. This involves recognising and identifying one's own emotional states. By paying attention to the bodily sensations associated with emotions, exploring the thoughts or beliefs that contribute to an emotion, and identifying the actions that tend to be associated with it, we can gain a deeper understanding of our emotions and learn how to manage them more effectively.

Another way to improve emotional awareness is through mindfulness practices, such as meditation or yoga. These practices can help individuals become more attuned to their emotional experiences and develop greater control over their thoughts and emotions. By cultivating a sense of calm and equanimity, we can learn to manage our emotions more effectively and respond to challenging situations in a more thoughtful and constructive way.

In addition to developing greater emotional awareness, there are other ways to improve emotional intelligence. For example, we can learn to empathise with others and understand their emotions. This involves putting ourselves in another person's shoes and imagining how they might feel in a given situation. By doing so, we can develop greater empathy and learn how to respond to others' emotions in a compassionate and supportive way.

Understanding and having awareness of one's emotions is crucial to developing a practice of intuition. The first step is to get a feel for how aware you are of your emotions. As we've discussed, this isn't as easy as looking at your heart rate on your smartwatch, although getting biofeedback like that can be helpful in learning to be aware of your emotional state.

There is a phone app called Mood Meter (with which I have no affiliation or connection), conceived by researchers at the Yale Center for Emotional Intelligence, which is based on the theory that building a habit of labelling and understanding how you feel at any given time will improve emotional intelligence. To be clear, I'm mentioning it here as a tool for visualising the spectrum of your feelings, as a step towards improving emotional awareness in relation to knowing when it's safe to practise intuition.

When you go into the app to report how you're feeling, the process begins with four quadrants, as in the following simplified representation:

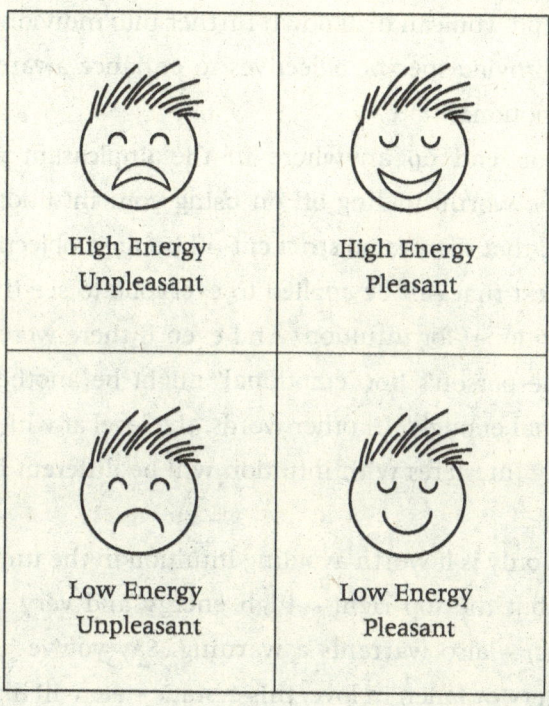

High Energy
Unpleasant

High Energy
Pleasant

Low Energy
Unpleasant

Low Energy
Pleasant

I think the spatial layout, with high energy at the top, low energy at the bottom, and a progression of pleasantness from left to right, provides a nice way to visualise the spectrum of emotions. The straightforward, four-part layout allows for quick identification of feelings. However, in the actual app, you can drill down further into individual cells, which provide specific adjectives to enhance awareness of your emotional state.

If you end up anywhere in the unpleasant zone, it might be worth holding off on using your intuition. Keep in mind that there is no strict cut-off point or objective universal test that can be applied to everyone to see if they're too emotional for intuition. And even if there were such a test, one person's 'too emotional' might be another's 'not emotional enough'. In other words, the level at which emotionality interferes with intuition will be different for each person.

Not only is it worth avoiding intuition in the unpleasant zones, but the top right – high energy and very pleasant (ecstatic) – also warrants a warning. Say you've just won the lottery or fallen in love, this ecstatic state will likely also interfere with your ability to feel the subtle pulls and nudges of your intuition.

If you've checked in with your emotional state and found you're stressed, anxious or depressed, then don't practise intuition. But what then?

When it comes to alleviating stress and anxiety, there are a range of things you can do by yourself at home, anytime.

For example, to induce an immediate change in state, try a breathing technique, such as box breathing. For this technique you count to five slowly during an inhale, hold for another count of five, exhale for five, hold for another count of five, then repeat. Non-sleep, deep-rest protocols like Yoga Nidra can be very effective for relaxing the body and mind.

Then there are a range of things that can help more generally. Exercise seems to be as beneficial a treatment for mild depression as antidepressants. I for one find a huge mood boost from using the sauna. Sleep, of course, is important, as is reducing alcohol and sugar intake. Getting enough exposure to sunlight, especially in the first half of the day when your body's circadian rhythm is reinforced, and engaging in social, in-person activities can help improve and maintain your wellbeing.

These recommendations are of course in no way intended as a substitute for medical advice in the case of serious mental challenges, they're just potentially helpful suggestions. Remember, your intuition and decision-making are generally at their best when you're in a slightly positive mood but not ecstatically, over-the-moon positive.

Mastery

s$\mathbf{M}$ile

Learn Before You Leap: Why Mastery Matters for Intuition

The magic of Hollywood learning

In the science-fiction film *The Matrix* the main character, Neo, is getting ready to confront dangerous agents in the matrix and needs to urgently learn years' worth of self-defence ASAP.

In one of my favourite scenes, Neo needs to learn jiujitsu. The operator spins up a computer screen with *jiujitsu* written on it. Neo looks sceptical. The operator winks at him, hits a few buttons on the screen and Neo recoils, scrunching up his face in shock. He rocks back and forth on his chair as if a magnitude 10 earthquake is tossing the room around. The skills of jiujitsu are being uploaded into his brain, which has been accessed via a sharp spike inserted into a type of computer plug at the back of his head.

Uploading a new skill directly into the brain via a computer plug without doing any actual learning: how good would it be? For anything at all that you want to learn, you just hit a button and there you go. Instantly you're a Grand Slam tennis player, a helicopter pilot, Albert Einstein.

Now I hate to be the bearer of bad news, but the brain simply doesn't work like that. The brain is not a computer. When we learn something new, our brains physically change. They rewire themselves, cutting older connections and building new ones. What's more, we all have very different brains and minds, just as we all have very different bodies, muscles and skeletons. To be a master tennis player, you would need to have your brain rewired in a bespoke manner, in a way that is unique to you, to your body, bones and muscles. And even if we could somehow restructure your brain to match that of Serena Williams, that still wouldn't work because what works for her brain and her body will be very different from what works for yours.

The same issues apply to uploading new knowledge and cognitive skills. Your brain is yours and yours alone, and its wiring is like your fingerprint – unique to you.

Perhaps the best explanation of how the brain is changed by learning comes from neuroscientist David Eagleman in his book *Livewired*. The brain is neither hardware nor software, but rather something he calls liveware – a dynamic system in which the so-called hardware and software mingle. They are in a delicate, continuous dance, shaping each other in amazingly elegant ways. Hence it's not enough to change only one of those components, say the hardware. You need to change the software, the brain's activity and your mind as well, and we are a long way off figuring all this out.

Why the 10,000-hour rule for learning is wrong

The late Kobe Bryant sits on a black stool off to one side of the TEDx stage at Shanghai Tower. Bryant, one of the all-time basketball greats, quietly and almost shyly says hi to the audience and thanks them for being there. He's doing a Q&A with the audience. A few questions in, the interviewer says, 'One moment that stands out to me was when we went out to practise at *four am*.' He emphasises the hour.

Bryant says, 'To me, it just makes complete sense. If your job is to be the best basketball player you can be, you have to practise. You have to train, you want to train as much as you can, as often as you can.' He goes on to detail how much more training he's done each day simply by starting at four in the morning, and adds, 'As the years go on, the separation you have from your peers and competitors just grows larger and larger and larger. By year five or six, it doesn't matter how much extra work they do in the summer, they're never going to catch up because they are five years behind.'

This deceptively simple logic of Bryant's has surely pumped up thousands of kids around the world, got them up and out the door early to out-practise the next person. But it's not really how learning works. It's not just a matter of the number of hours you put in – some hours are worth more than others.

Unlike, say, walking, learning isn't a linear process. With each step I take, I'm a metre closer to my destination.

But each hour of practice I do doesn't get me one hour closer to my goal of being an NBA star. It certainly helps, but the value of each hour depends on a whole range of factors. In fact, as we'll see, sometimes you can master something in a single hour.

In his 2008 book *Outliers*, Malcolm Gladwell promoted the idea that anyone can become an expert at something with 10,000 hours of practice. He convincingly walks the reader through examples of the first twenty years of people's lives. Those who went professional in their field had around 10,000 hours of practice under their belt by that point, while those who didn't make it had only 2,000 hours. He goes on to say that 'researchers have settled on what they believe is the magic number for true expertise: 10,000 hours'.

This idea, perhaps not surprisingly, was very sticky and went on to cement itself in the minds of many as a true number for expertise. But in fact, it's inaccurate, and not a good way to think about learning something new.

Repetition, the number of errors you make, and the motivational outcomes of learning are what matter. The context around your learning – for example, how much sleep you get – also matters. Learning is not about the sheer number of hours you put in, it's about the quality of those hours, and about what you do after the learning, which can help your brain make that learning permanent.

Tasting the invisible dog food

The type of learning that drives intuition is known as associative learning or classical conditioning. It's a little different to the type you might be more familiar with, such as learning the piano or skiing.

Towards the end of the 1800s, the Russian scientist Ivan Pavlov ran experiments on classical conditioning in dogs, focusing on their salivary, gastric and pancreatic secretions. Pavlov's dogs became famous as a result, and Pavlov himself became the poster child for classical conditioning – the process by which the brain learns that one thing predicts another. When dogs are about to eat, they produce huge amounts of saliva (in Pavlov's time, these gastric juices from dogs' mouths were actually a popular treatment for indigestion in people). The experiments Pavlov and his staff ran with these gastric juices were not the nicest, so I'll spare you their details, but Pavlov noticed something interesting.

At the start of the experiments, the dogs would salivate when food was put in front of them. But over time, they began to salivate before the food arrived. The now-famous story is that the staff would ring a bell when it was time to feed the dogs, and after a few days of this, the dogs began to salivate at the sound of the bell alone. Pavlov called this salivation 'psychic secretions', because the dogs were behaving just as if the food were actually in front of them.

But of course, it wasn't. The sound of the bell was now producing the same effect as the food itself, and he called this a 'conditional response'. The dogs' salivating was conditional on the animal learning that the bell predicted the food. Pavlov went on to receive the 1904 Nobel Prize for his work, which is still discussed in every psychology course today.

This conditional response has become known as Pavlovian or classical conditioning. It remains the basis for our understanding of learning and psychological disorders and their therapies. It's also why you get flooded with emotion and memories every time that particular song comes on the radio, and why your cat runs to you when you grab the can opener. At the heart of it is a very simple idea – that when neurons fire together in the brain (or closely in sequence), they get wired together. They literally get connected, via the small branches that go from one neuron to another. These connections mean that things you see or hear one after the other in the world become associated with each other. The bell rang and the dogs started to salivate because the food neurons in the dog's brain were linked to the bell neurons. The bell activated the food neurons.

This associative learning is very different to a dog *deciding* to think about its food when the bell rings. Associative learning connects different parts of the brain, so the process of salivating becomes automatic. Associative learning occurs even in very small and simple animals like sea slugs. It can occur unconsciously.

In fact, the automatic learning of associations is the backbone of intuition. It's how intuition develops from your everyday experiences. Each time you eat in a café and have a good experience, for example, the neurons in your brain that represent the café and the neurons that represent the happiness of the experience become slightly more interconnected. And each time you eat at a café and have a bad experience or get sick, different networks of neurons become wired together.

As a result, you begin to feel something when you go into a café. Positive or negative feelings are triggered without you having to do anything. It all runs automatically, like Pavlov's dogs and the bells. The neurons processing all the details in the café – the tablecloths, the music, temperature, smell, etc – trigger the emotional parts of your brain, and you feel something. This is the sense that people describe as coming from the gut. The interoceptive gut feeling of intuition. It is what saved Jon Muir's life on Everest that windy morning on the western ridge. The neurons in his brain that processed the wind, temperature and light triggered emotional neurons in his brain, and he felt the activity of these brain areas as emotion in his body, in his gut.

If you have no experience with something – maybe it's your first time climbing, or playing tennis or chess, or making investment decisions – then you won't have any such useful associations and therefore cannot rely on intuition. This is why experience and mastery of something is a must if you are going to trust your intuition. You cannot rely on intuition if it's the first time you're doing something.

How much mastery is enough for intuition?

Unfortunately, there's no simple answer to this question. As we saw with acquiring a skill, where it's not just the number of hours spent on practice that matters, but the nature of those hours, mastery is similarly dependent on the nature of the experience.

Negative outcomes have the power to accelerate learning. A good rule of thumb is that the more negative the outcome, the less experience you, and your brain, need to learn something. Consider post-traumatic stress disorder, PTSD. Many people who experience trauma from a single event, such as a car crash, learn extremely strongly from it. So strongly, in fact, that it becomes problematic.

But less-extreme negative experiences can also work like this. In my own case, the event was a cricket match – the first one I'd been to and, as it turned out, the last. It was also the first time I'd drunk rum. My friends were buying the drinks that evening, at a pretty regular pace, and I've never been much of a drinker. At some point, the rum having overpowered me, I disappeared to the bathroom. From where I emerged hours later, dazed and confused. The place was empty, completely quiet, the crowds gone. It was late at night, but the arena was still brightly lit. I had no idea where my friends were.

That was the last time I ever drank dark rum. That one evening not only put me off cricket for the next twenty-five years and counting, but also rum. If I so much as smell dark rum today, I am transported right back to that evening and the feeling of nausea. I became an expert in not drinking rum in a single evening.

This is an example of the brain's power to rapidly learn that something will make you sick and keep that memory for a lifetime. It's an example of associative learning. I didn't try to remember that rum can make me sick; in fact, I would have preferred to forget it. But my brain linked the two things, rum and being sick, together forever. Possibly cricket too – I've never tested that one.

There's an interesting asymmetry between positive and negative experiences. Negative outcomes are almost always more powerful and affect us more than positive outcomes. Hence, when it comes to the associative learning behind intuition, negative things have more power to drive learning. It makes sense that our brains are like this, since negative things are often life-threatening, and we need to rapidly learn from any encounters with them. There are exceptions to this, of course, such as sex and drugs, which can feel positive in the beginning and then turn into addictions. While sex generally goes on to remain positive, it too can become an addiction.

What all this means is that how much learning you need to build a basis for intuition depends on whether your experiences are positive or negative and how strong those experiences are. A single instance of something, when the results are strong enough, can be all you need, as it was for me and the rum. For other things you might need a thousand repetitions of an event, or two thousand, to gain enough experience to form the required associations to drive intuition.

There's one more factor in the mastery–intuition equation. The environment also has an effect on associative learning. Environment, you'll recall, is the fifth rule of SMILE, and we'll come to it in detail later. Suffice to say here that the more things there are around you that can predict the outcome, the thinner the associations get spread, and the weaker the learning is. Hence, busy or cluttered environments require more iterations of the learning process because there are more predictors for your brain to learn about. The detailed nuances of learning theory are too complex to go into in this book, but in short, the strength of the outcomes is the primary driver when it comes to how much experience you need to trust your intuition.

Why timing matters

We've all been there. After a fantastic meal out with great company, you go home, maybe watch or read something before getting ready for bed. You go through your sleep routines and readily fall fast asleep. Then in the middle of the night, disaster strikes. A churning, gurgling sensation in your stomach snaps you wide awake in a second.

Food poisoning is never fun, and it can turn you off a restaurant, a particular food, even the specific flavours in the food. But here's the interesting thing: this food aversion, a type of associative learning, happens even though hours may have elapsed between your eating the food and sickness kicking in. Remember that with this type of learning you don't have to consciously think, 'Oh yeah, it must have been that fish I had earlier'. Your brain, without you having to do anything, learns the link between food and sickness.

Even more interestingly, taste aversion can be induced. You might have eaten a lovely dinner, after which I give you a pill that makes you really nauseous. Yes, I know, what a nice thing to do. What's more, I tell you that the pill will make you feel sick, so you're fully aware of the cause of your nausea. But your brain will still link the food you had for dinner with this sickness, despite you knowing full well that it wasn't the food.

A strong example of this is when cancer patients are given chemotherapy drugs for treatment and are fully aware that the drugs cause nausea: often, foods that are eaten around this time will result in an aversion – the patients begin to dislike the food. A similar but weaker phenomenon can happen when you have the flu. Hence it's good to avoid your favourite foods when you are sick.

However, most often with associative learning the connection happens fairly quickly. You go into a café, order a coffee and it's really bad. That feedback, getting a bad coffee, happens in only a few minutes. In many of the laboratory science experiments on associative learning, the feedback is almost immediate. The rat or university student makes a choice and gets a reward or punishment right away.

When it comes to the role of associative learning in intuition, sometimes the learning is immediate – as in many of the science experiments focused only on learning – but sometimes it's not. When playing football or basketball, an intuitive choice to go left instead of right will induce immediate feedback that it was either a good or bad choice. However, learning intuition is often more like the food poisoning example. Many hours, or even days or weeks may pass before you really get the feedback about your choice. Think about choosing to trust someone, making an investment choice, or deciding whether to buy that house you've been looking at. With decisions like these, the feedback from your choice, which your brain needs to learn from, can take a long time.

The thing with learning intuition from these longer feedback choices is that associative learning tends to be weaker the longer the delay between the predictors and the outcomes. Keep in mind that learning is also faster with more negative outcomes, such as getting sick from food poisoning. So, to bring these three rules of learning together, choices with immediate, strong negative feedback drive the strongest learning. Choices with delayed, weak feedback will only induce learning slowly.

If you are trying to develop your intuition by choosing which stocks to buy but you have to wait months to see how the stocks perform, any learning will be very weak, unless you lose everything and it really hurts. Otherwise, you'll need many interactions to develop intuition. The inverse is also true – for choices and actions with immediate feedback, the learning here is much stronger, so you won't need as much repetition to learn intuition.

When it comes to making intuitive choices or actions, the timing matters. Short gaps between your choices or actions and their outcomes will induce stronger learning, compared to choices with a long wait for the feedback. The additional bonus of short waits for feedback is that they enable you to practise more choices with feedback in the same amount of time. So if you want to optimise building your intuition practice, start with things that have tight feedback loops.

Protocols for maximising learning in intuition

Research shows that failure and making errors are crucial to learning anything. They send a shock signal through your brain, telling it something didn't work as expected and therefore needs to change. This nudges your brain into a plastic, or malleable, state. It gets ready to change itself – the perfect state to be in for learning something new.

This phenomenon gets even more interesting when you make a lot of errors learning to do one thing and then switch over to learning something else. Your brain is already in a changeable state, and in a sense you can trick it into learning the second thing by making errors at the first thing.

The problem is that most of us don't enjoy failure. We don't enjoy getting it wrong, throwing darts into the wall, falling off a bike, whatever. Failure typically makes us feel that we're no good and that a lot of work is needed to become good at something. Obviously no one wants to fail 100 per cent of the time, that's not going to work either. But there is a sweet spot for a failure rate, somewhere around 20 per cent. The trick is to make failing more enjoyable, make it become more of a game.

This is in fact exactly what video games do. They use positive reinforcement and rewards in very particular ways to boost dopamine in just the right way, so that you start to find failures rewarding. You can actually look forward to them because they tell you how to make it to the next level of the game. The point is that once the sting has gone from screwing things up, your progressive learning can really take off.

These games also employ a range of other tricks, such as variable reward schedules, and the uncertainty of when you will win, in a similar way that a poker machine does, and with similar strategies to social media platforms. Of course, these strategies aren't always good, as the following incident from an online gaming café in Taiwan shows.

Spread out across many rooms in this café, hundreds of people are plugged into computers, ears sealed off from the world with hi-tech headphones. They're deeply engrossed in World of Warcraft, Final Fantasy, Fortnite, League of Legends, and other games. Long lines of red, couch-like chairs prop the gamers up against the long rows of desks. Hands hover over customised multicoloured gaming keyboards. The lighting is bright, not full-daylight bright, but enough to keep you awake 24/7. The clicking of keyboards and mouses sounds like a school of cyborg cicadas, disrupted occasionally by angry shouts of frustration and sudden screams of victory.

When the local Taiwanese police enter in a stream of uniforms, the sea of players don't even look up, they're oblivious to what would normally command plenty of attention. The clerk at the internet café shows the police to a corner where a young man sits stiffly propped against the desk. His screen monitor is still on, illuminating his face with blue and white light. One of the squad checks the young man, shakes his head and speaks quietly to the others, who nod solemnly in agreement. The officers pull out some yellow police tape and cordon off the corner where the young man sits wedged in place.

A few gamers close by look up, then just as quickly turn back to their monitors and games. The police squad look around in amazement, not trying to hide their shock as the gamers show no concern that one of their own is dead in his chair. He has in fact been dead for a while now, frozen motionless in place for hours.

It was later reported that the man, Hsieh, had died of a cardiac event many hours prior; it took ten hours for a clerk to notice that he wasn't moving and to check in on him. Hsieh had been in the café for over three days straight. He hadn't left the café at all during that time.

These gaming cafés are open twenty-four hours a day and provide food and drink to their customers while they play. Cold temperatures and over-exhaustion from the long hours likely contributed to Hsieh's cardiac arrest, a police statement said.

A staff member was reported as saying that Hsieh was a regular in the café and always played for consecutive days. He would sleep face down on the desk when tired, or slumped in his chair. This was the reason it took so long to realise that he was dead.

Stints of multi-day online gaming are known as Gaming Disorder. The World Health Organization has characterised this disorder as 'impaired control over gaming, increasing priority given to gaming over other activities to the extent that gaming takes precedence over other interests and daily activities, and continuation or escalation of gaming despite the occurrence of negative consequences'. Death during gaming also has a name – sudden gamer death – and is more common than anyone would like.

Just why are these games so mesmerising and addictive? There's clearly something different about learning to win a game compared to other types of learning. When most of us sit down to learn something new, we don't get immediately addicted to the process. When was the last time you heard about a high school student who couldn't stop studying for three days straight?

The answer lies in the process games use to make failure exciting and rewarding. All of them involve very challenging activities that require focused attention, long hours of learning, and hundreds of thousands of painful setbacks and failures; players 'die' over and over again. But in gaming these failures are nevertheless rewarding. Can we harness this process to boost our learning for intuition?

Learning can be sped up not only by making errors, but by paying attention to the source of your errors, how and why you got things wrong. As we've seen, the type of learning behind intuition is associative learning, which can occur without you being conscious of it. However, paying attention to the things in the environment that immediately contributed to your errors and the outcome of a decision will boost learning.

Surprise is the other factor. When you get something wrong that you were confident about, you not only make an error, but you're also invariably surprised by the mistake. These surprise errors grab your attention and immediately focus it on the error, thereby clearly informing your brain what it needs to change. Surprise is like a highlighter marker – it lights up the important items for your brain to boost learning for, and avoid subsequent surprises.

An important part of learning actually happens after you've stopped actively doing or practising the thing. Then the brain begins a different process called learning consolidation. Think of this as your brain locking in what it has learnt into long-term storage. This process begins almost straight-away, but also continues through your next night's sleep. Thus, not only is getting a good night's sleep important, but short rests, even naps, right after learning something new can help boost your learning. I'm not saying you *have* to take a nap every time you learn something new, but it is good to be aware of the concept of learning consolidation and why making space for it after practising something can help.

We've seen how timing matters with learning. In addition to this, tracking your learning can be a powerful way to see progress and boost motivation, attention, and hence learning, in a similar manner to how video games tend to show progress with points, money, or some other unit of progression.

One final point regarding learning is that you can also overdo it. More is not always better. In fact, data shows that practising something continuously for too long can actually harm your learning. As you become tired, your practice can get sloppy, and your technique with it. The key is understanding that your cognitive and attentional resources, and of course your physical resources, are finite. Over time, without a break, the efficiency of learning decreases. For example, you might find that during the first half-hour of practice, you're absorbing a lot and making fast progress, but as you push on, the returns diminish.

It's challenging to pinpoint a specific 'golden' duration for practice that applies universally to everyone, because learning is influenced by many factors, such as the nature of what you're learning, your motivation, cognitive and physical capacity, prior experience, and more. A general approach is to start with a shorter duration, say twenty minutes, or something that seems reasonable to you, and if that works, increase to thirty, then forty or even fifty minutes.

How to know when you're a master

Magnus Carlsen, the Norwegian chess grandmaster, is one of those prodigies you picture with the mention of chess masters, those people who live and breathe the game and have it soaked into every cell of their bodies. If anyone is going to have mastered something, it's safe to say Magnus has mastered chess. So, what actually gives him prowess? Let's take a look at an experiment he took part in.

He sits down across a chessboard from David Howell, an English grandmaster and commentator. They're going to play parts of historical games as a test. First up is the 1960 world championship match: Mikhail Tal versus Mikhail Botvinnik. On the board in front of Magnus and David is the exact layout of each piece from a moment in that game. Magnus looks down and without any hesitation says, 'Aha, yes, this looks an awful lot like Tal–Botvinnik and I think the continuation here was probably queen D5 and then rook A6 and Tal won.'

'Yes,' says David, nodding and smiling. Not only did Magnus recognise the match immediately, he knew what the next few moves were.

Next, David starts rearranging the board to show a snapshot from a 1987 world championship game: Garry Kasparov versus Anatoly Karpov. But before he has finished moving the pieces Magnus says, 'This is the twenty-fourth game from Seville, obviously.' David laughs out loud and drops the piece in his hand. 'You got that right, shall we move on?'

David decides to try something different. 'Now I'm going to play through an opening, and stop me when you recognise the game and tell me who was playing black.' He sets the board up in the formation of the game, then moves one of the white pawns, the smallest of the chess pieces that fill up the second line. Then he moves the mirroring black pawn on the other side. The next move is a white knight, followed by a black knight. At that point Magnus says, 'Okay, it's going to be Anand.' David's disbelief and amazement is apparent. He has moved just four pieces.

'Against?' he asks Magnus.

'Zapata,' says Magnus, smiling.

'What year?'

'Ah, '87, '88,' says Magnus.

'Yes, '88,' says David.

Now David moves on to the trick test, a fictional game from the first Harry Potter film, in 2001, in which Harry is playing against enchanted chess pieces. Once David has set up the board, Magnus stares at it. 'This could have been some Bold game,' he says, referring to BoldChess, an online academy for learning and playing chess.

'That's a good guess,' says David.

'Hmm. I'm probably going to need a hint,' says Magnus.

When David tells him it's from the entertainment industry, Magnus says, 'Okay, black is down a queen, so from the first Harry Potter movie.'

Does this feat mean that chess players have a better memory than everyone else? No. It turns out that if you put the chess pieces in random positions all over the board and ask chess players and non-chess players to try to remember the layouts, then the chess players don't perform any better than non-chess players.

The way memory works is that as we get more familiar with things, it gets easier to group them into patterns or create larger compound objects that are easier to remember. Short-term memory is severely limited in most of us. We can only hold between four to seven items in it for short periods. The trick is to find a way to arrange a collection of items into one coherent pattern or object. Then, instead of having to remember, say, twelve items, those twelve items become one, and your other memory slots are free again.

This is exactly what's going on with chess experts. They're not remembering the location of each piece on the board and rehearsing those locations over and over until they stick. They're remembering the Smith opening sequence, or the Jones move set – layouts on the board that are recalled as a single item. This method is called chunking, and it leads to a massive increase in what can be remembered. It works for short- and long-term memory.

For long-term memory, another technique you might have heard of is a memory palace. This is the method of imagining a place, say your house, in your mind's eye and placing all the objects you want to remember at different locations as you walk through the house. The extra step here is to make all the items memorable and meaningful, so they're easy to remember. You can learn to have a memory like Magnus by practising these simple techniques.

How does this relate to knowing if you have enough mastery for intuition? Learning is a form of memory; as your brain learns all the associations between the predictors in the environment and probable outcomes, this is stored in a form of memory. However, as we've discussed, it's a different type of memory than simply storing facts about how to move chess pieces. Magnus probably began to know that he was a chess master when he started winning all his games. How can you know when you have mastered enough learnings to use and trust your intuition?

Unfortunately, as with the question of how much mastery is enough for intuition, there are no precise signs for knowing when you have enough. But a good way to know if you're well on your way is to monitor the frequency of surprises. As your learning progresses and plateaus out, there should be fewer surprises.

Importantly, as you practise intuition, make sure that all five SMILE rules are met. Get really familiar with them so that they become second nature. Whenever the rules are met, practise your intuition and note how successful it was. For example, when you made an intuitive choice between two cafés and were not surprised by regret or pleasure from that choice, this suggests that your intuition might be working. In other words, you had no surprises from following your intuition.

Of course, a one-off example like this won't by itself provide much evidence of mastery, but over time, with many instances of using intuition without surprise, the evidence can start to stack up. As we will discuss later, it's a good idea to begin practising intuition with decisions that have small impacts. Don't start off with life-changing decisions.

It's also important to keep in mind that your path to mastery lies through associative learning. Associative learning is context-specific, so developing it at work won't generalise to your home environment. You will need to develop your intuition in each area you want to practise it in. For example, practise in a work context to develop work-specific intuition, practise at home for home-specific intuition, in a sports context for sports intuition, and so on.

As you practise the five rules for intuition until they become familiar, start noticing the interoceptive feeling of your choice or action. Follow it, then assess the results to monitor the level of success. The whole process will become more and more automatic. This is how learning works. Do you remember the first time you drove a car? Each small movement would have required lots of focus and attention; you probably hit the brakes too hard and fast, causing the car to lurch to a sudden, uncomfortable stop. Hundreds of hours of driving later, you might find yourself arriving home without even paying attention. All those different movements once so difficult are now so easy, like walking.

This is the amazing power of your brain. As we learn new skills the experience of doing them radically changes, they become automatic and even unconscious. This also applies to using your intuition.

Impulses and Addiction

SM**I**LE

Never Mistake Impulsive Desires for Intuitive Insights

Instincts are not intuition

Cast your mind back to early 2020, when the COVID pandemic was just kicking off. The virus plunged everyone into a whirlwind of uncertainty. Would we all get really sick? Die? Would everything shut down? Australia was already on edge, having just lived through a harrowing summer of devastating bushfires along the east coast. We just didn't know what was going to happen next.

It was uncertainty that led to those barren supermarket shelves once filled with toilet paper or pasta. Then videos started popping up online of people brawling over the last rolls of toilet paper. Pulling clothes, hitting, kicking, screaming. They looked like a *Fight Club* re-enactment, or a drunken New Year's Eve brawl.

What next? No food? No water or power? My wife, having grown up in Venezuela and, unlike me, lived through shortages of food, medicine and electricity, takes this kind of thing very seriously. I, on the other hand, didn't feel seriously threatened by such possibilities. This is the thing about uncertainty – it affects different people differently. Some people are really hit hard by not knowing what will happen next, or not seeing clarity in the present situation, and being engulfed by ambiguity drives the fear centres of the brain. In fact, we primates are hardwired to fear uncertainty.

Because this innate dread isn't learnt from our environment, it can, at times, be maladaptive. Our brains react to uncertainty as if we've just stumbled on a venomous snake or spider. Despite the individual differences with uncertainty, the thought of an unknown future scares the living daylights out of most of us. Psychological experiments have shown that we even find looking at blurry images uncomfortable. Take a photo, blur it in Photoshop, and most people will rate the blurry image as uncomfortable. People don't like looking at ambiguous, blurry things. This innate fear of uncertainty wreaks havoc on our decision-making abilities in all sorts of ways – think risk aversion or decision paralysis.

The modern world is no less filled with uncertainty than was that of the first *Homo sapiens*. New viruses, climate change, war, AI – this built-in fear of the unknown makes life really tough to live. Because of our dislike of uncertainty, we'd rather, for instance, wait longer for a delayed flight with a clear, rescheduled time than wait a shorter period with no idea how long the wait will be. We find the shorter wait more painful because of the uncertainty. Uber's success is often attributed to its innovative approach to reducing travel uncertainty. Before Uber, you'd stand on the corner wondering when a taxi would show up, how much it would cost and how long it would take. Uber stripped away the uncertainty, and with it the fear and discomfort.

INTUITION

In a world dominated by technology, our innate instincts and reflexes struggle to navigate the uncertainty woven into our daily lives. When faced with it, we often revert to our primitive selves, freezing in fear, feeling anxious, and panic-buying supermarket items.

But what if I told you that embracing uncertainty can be a powerful advantage in the modern world?

The first step in thriving in uncertainty is to understand that the fear you might be feeling is not adaptive, it's not your intuition telling you there's danger in your environment. This fear can be better characterised as a leftover product of evolution, from times when uncertainty often meant the difference between life and death. Once you understand this, there are practical guides to learning to move through this fear, to condition your nervous system to be okay with it.

Achieving this can open the door to not only a better quality of life and improved wellbeing, but also an increased level of creativity and, yes, even intuition. The fear you feel of the unknown is not your intuition telling you something is wrong, it's an instinct, and instincts are innate. Intuition is learnt and adaptable. It adapts to your environment.

Impulses and Addiction

The 'I' in SMILE is for impulses and addiction: impulses are innate reflexes we come into the world with. They are not learnt, like intuition is, they are intrinsic, and we shouldn't confuse our instincts for intuition. In the animal kingdom, instinct is king. Salmon swim upstream to spawn. Birds migrate thousands of miles guided by unseen forces. Newborn turtles instinctively crawl towards the sea moments after hatching and return to the exact spot years later. But in the complex world of human beings, the lines between instinct, reflex and intuition get blurred. To navigate this intricate labyrinth, we must understand the distinctions and the roles they play in our lives.

First, instinct. That sideways twist of the mouth and squinting of eyes on biting into a lemon is an instinct, present even in babies. We come into the world pre-programmed with responses to certain flavours and smells; these responses help keep us alive by stopping us from eating poisonous things.

Reflexes are the body's automatic responses to stimuli, designed to protect us from harm. These are simple, rapid reactions hardwired into our nervous systems, such as pulling your hand away from a hot surface, sneezing when irritated by particles in the nose, or the classic knee-jerk reaction when tapped below the kneecap. Reflexes occur without conscious thought and are crucial for our survival in a world fraught with potential hazards – but they are not intuition.

But while instinct and reflexes are genetically determined and have evolved over generations to be part of our DNA, they're often ill-suited for the modern world. The world is changing very quickly, but our instincts don't; they are hard to change. Intuition, on the other hand, being easily adaptable, is tailor-made for our contemporary lives. While our capacity for it is innate, its specific application to decisions is learnt from experience. It learns on the fly, constantly adjusting itself to our changing environment.

The problem arises when we confuse the two. Instinct and intuition can feel the same, so it's easy to mistake one for the other – many articles and books, in fact, use the two words interchangeably. But our instincts and reflexes can get woefully out of date, and it's important to remember this fact.

Another example of out-of-date instincts that we shouldn't confuse with intuition is our innate drive towards comfort. Imagine, for a moment, that you're an early human ancestor midway through a brutal winter. You're shivering, teeth chattering, as you trudge through snow-covered plains in search of your next meal. The biting cold numbs your toes, and each breath feels like you're inhaling shards of glass. Fast forward to today, and most of us are comfortably sipping lattes in climate-controlled environments, grumbling when the wi-fi signal isn't quite fast enough for our 4K TV.

Don't get me wrong, central heating is a marvel, and I'm not suggesting we toss out our thermostats and revert to Flintstone-style living. But our bodies, refined over millennia, are built to handle those brisk ice-age dawns. We possess a mechanism called thermogenesis – a built-in furnace of sorts – that kicks into gear when faced with cold; our bodies also produce health-enhancing heat-shock proteins when it gets blisteringly hot. Yet, due to our innate drive for comfort, we've outsourced these bodily responses to technology.

In a way, we're shelling out cash to make ourselves more fragile. Our fixation with comfort means we're bypassing the health advantages of both chilling out and heating up. Exposure to cold has been shown to elevate mood, bolster the immune system, and even assist in burning fat. On the flip side, millions are reaping the rewards of regular sauna sessions, with science backing up the numerous benefits of experiencing both temperature extremes.

So why is it that we shy away from discomfort? Our affinity for comfort is deeply ingrained – an instinct we're born with. And undoubtedly, in bygone eras, this instinct proved lifesaving. But in today's world of abundance and convenience, an overdose of comfort has transformed this once protective impulse into something maladaptive. So the next time you're faced with uncertainty or discomfort, remember that embracing the unknown, and sitting with the feeling of discomfort, rather than blindly following your instincts, can be powerful weapons in today's ever-changing world.

Learning to thrive in uncertainty, to do hard things that are uncomfortable and to live without fear in their midst can give a strategic advantage in the modern world. By distinguishing between your instincts, reflexes and intuition, you can better adapt and thrive in an age dominated by technology and rapid transformation – just remember that instinct is not intuition.

Un-intuitive eating

It's late at night and you're sprawled on the couch, when that familiar urge hits. It's as powerful as a hurricane and there's no way you can resist it. You haul yourself off the couch and stagger to the kitchen, and before you know it you've got the ice cream in hand. Screw the bowl – you dive right into the tub because life's too short to waste time on such trivialities. You marvel at the elegant curl of the chocolate swirls that your spoon so effortlessly carves up, and then, the flavour! The cold texture melts on your tongue like the world's most delicious snowflake. You don't bother to pace yourself, digging in for spoonful after spoonful, skilfully carving a prize-winning ice sculpture.

A voice in your head tells you to stop, but you silence it with another mouthful. Your body knows best, right? This urge is here for a reason. So you keep eating, revelling in the unadulterated satisfaction that comes from giving in. Before you know it, the tub's empty and you've sprawled out once again, feeling full and satisfied.

Half an hour passes and you feel a new hunger stir. Back for round two? Hell yes. Chocolate-chip cookies this time. You've given yourself unconditional permission to eat whatever your body urges, and you intend to do just that.

There's a whole movement called intuitive eating, which proposes that you honour your hunger and stop worrying about eating too much. It suggests that you make peace with food and give yourself unrestricted permission to eat whatever you feel like.

And eating that ice cream and the cookies feels so damn right, doesn't it? That internal craving is natural, and following it feels like the right thing to do. You've honoured your hunger and made peace with the food. You've banished any voices in your head saying, 'This is bad, don't eat it'.

But these cravings just keep coming, and they're growing stronger each time. Is this really intuitive eating? With modern food engineered to be addictive, how can this be a good idea?

These primal cravings remind you of the irresistible urge to check your phone and scroll through social media. The automatic, zombie-like walk to the kitchen becomes the robotic reach for your phone. Before you know it, your eyes are glued to the screen, thumb flicking up, devouring the endless feed of glossy images. It feels right, natural and intuitive – just like when you gave in to your food cravings – and you give yourself unconditional permission to scroll and scroll.

The truth is that powerful, irresistible cravings are an illusion of intuition. Many substances and behaviours, including alcohol and obsessive social media use, can create these cravings, tapping into our brain's reward system. The frightening reality is that they can lead to addiction, making it feel like nothing else apart from their consumption matters. They bring us feelings of pleasure and literally change the chemistry and wiring of our brains. And confusing these urges for modern food, cigarettes, alcohol or social media with intuition can lead to addiction. Cravings for these things are not intuition, and it's crucial to recognise the difference.

Intuition is a subtle feeling often derived from years of experience, shaping the associations between our environment and probable outcomes. Although cravings and intuition may at times feel similar, they are fundamentally different.

Cravings, addiction and intuition

Our cravings, just like intuition, originate from within us, from our internal perceptual system, the process of interoception. That's right, the system that governs our cravings, that navigates the labyrinth of addiction, shares its architecture with intuition. As counterintuitive as it may sound, the infrastructure that enables us to make split-second decisions is closely related to the one that leaves us yearning for another bite of that cookie. They both rely on learning.

However, the learning system can be hijacked by very pleasurable experiences, such as sweet food, nicotine, the endorphins we get from exercise, cocaine, sex. Enter dopamine. Dopamine is a crucial part of our brain's operations – a chemical messenger that helps tell our cells, and us, what to do. Dopamine is different to endorphins; it's the cool kid in the class of neurotransmitters with the knack for getting things done, especially when it comes to the business of hedonism and motivation. Dopamine is the driving force in eating that slice of cake, achieving the victory in a game, in the anticipation of meeting up with a friend. It's the foot pressing down on the accelerator, keeping us moving towards these enticing experiences.

While food and sex can boost your dopamine between 150 and 200 per cent, drugs like cocaine can boost it 500 per cent and amphetamines 1000 per cent. (Of course, these drugs are also doing many other things to your brain that can drive addiction beyond the dopamine spike.) The effects of these types of pleasurable experiences on the brain are so far beyond those in the normal learning process of intuition that they are simply not on the same playing field. Therefore, they're not directly applicable for our mission to build intuition and improve our lives by making better decisions.

The cravings that inevitably arise from these high-reward experiences are not the same as the positive sensations you feel in your body as part of intuition – or, indeed, the negative sensations. Like intuition, cravings are largely the product of your brain's learning systems. But, in comparison to intuition, the sensations of cravings are typically extreme – I'm talking Red Bull, jumping-out-of-planes extreme, diving-with-sharks extreme.

What's more, cravings are extremely positive, or at least tend to feel that way. The pull they exert on you is very strong, you will crave more and more and more. Think of these high-reward behaviours as intuition on steroids.

And just as with steroids, there can be nasty side effects. Your brain gets used to the dopamine hit provided by these experiences and they become the new normal. Hence, you get more and more dependent on them: a process called addiction. However natural and important the cravings may feel, they get stronger and stronger, leaving you needing more and more. Fulfilling them won't lead to better decisions and a better life. On the path of addiction, the brain is rewired to crave the constant hit of dopamine and other chemicals that these activities provide. And with each hit, the brain's natural ability to produce dopamine diminishes, leaving you feeling dull, uninspired and unmotivated.

That urge to grab your phone and check that social media feed for a new message one more time – is that intuition speaking? Sometimes, even the logic behind some of these cravings can seem like intuition. Maybe your brain has picked up on the subtle clues from prior posts and is predicting your post will go viral. You better check, right? How do you know if that's intuition or the craving for an addictive dopamine hit?

The key to knowing lies in the strength of the craving, because the strongest feelings in intuition tend to be negative. Jon Muir's sinking, sick feeling on Everest, for example. Cravings for addictive things almost always show up as a positive pull towards something.

Brains can be funny sometimes, and as we've seen, when it comes to positive versus negative emotions, the response to negative things tends to be stronger than the response to positive. It's simply the way the brain works. It's evolved over long periods to respond to negative, potentially dangerous, things more strongly because doing this is more likely to keep you alive. Missing out on picking a flower or patting a puppy might be a little sad, but missing a snake coming at you could end your life. This means that the positives and negatives of intuition are unbalanced. Negative things drive learning more, so you will feel them more often and more strongly as part of intuition.

Cravings and addiction march to a different drumbeat. Here, it's not about being pushed away by potent negative emotions, but about being lured by a strong positive attraction.

So next time you're confronted with a compelling desire for something, when you find yourself yearning for it, when it feels as if attaining it would be a major win, as if it's all that matters – hit the pause button. Take a moment to check in with yourself. It's quite possible that what you're feeling is not the subtle nudge of intuition, but the pull of something with addictive potential. A siren's call cloaked in the guise of positivity, leading not towards the safe shores of intuition, but to the treacherous rocks of addiction.

However, a few exceptions to this rule are worth mentioning. The first is love, which in its unadulterated form can evoke powerful, authentic, deeply natural feelings. An overwhelming gravitational pull towards someone can sweep you off your feet, and it's plausible to think that intuition plays a role in this magnetic attraction. Therefore, the rule of thumb laid out above may not apply in the same way when it comes to matters of the heart, such is the complexity and nuance of our emotional landscape. Sex, too, can be addictive: sex addiction is an actual condition. But I'm not suggesting you stay away from sex, so let's add that as another exception to the rule as well.

The third exception is physical activity. Many people who exercise regularly begin to crave it. This is often related to the release of endorphins: peptides in the brain that act as natural painkillers and mood elevators. Craving exercise can be a healthy habit that promotes physical wellbeing, but it can also be addictive, a condition known as exercise addiction or compulsive exercise. It's characterised by an obsession with physical fitness and exercise, often to the detriment of health and wellbeing. It's relatively rare, but it can be serious.

A fourth exception is social interaction. Humans are social creatures, and many people have a healthy craving for social interaction. This is another pull that doesn't obey the rule of thumb, since social connection can greatly contribute to emotional wellbeing. In addition, many people who work in creative disciplines describe their relationship with being creative as being addictive. Visual artists talk about the trancelike state they enter when painting, sculpting or drawing. This immersion in work can be likened to an addiction due to its intense and sometimes overwhelming pull.

To recap: just as instinct can feel like intuition, so too can impulse and addiction, but these things are not interchangeable. You need to be on the lookout for cravings that might slip under the radar as intuition, and ensure that you don't justify giving in to your cravings by calling them intuitive. Learn the differences and how to spot them in your practice of intuition.

Broken compass: don't use intuition if you have an addiction

Research has shown that people with substance addiction, or a behavioural addiction like gambling, show clear differences in their decision-making. Let's take the example of Rob Ford, who was mayor of the city of Toronto from 2010 to 2014. Ford seemingly had a rough few months at the end of 2013, attested to by several video clips.

One shows him smoking crack cocaine through a tube. Another shows him addressing a group of journalists, during which he admits to having smoked crack but denies being an addict. In yet another grainy, hand-held video, he is dressed in a shirt and tie, in what looks like a meeting room, energetically pounding his fist into the air as if punching an imaginary person. 'I'll rip his fucking throat out ... I'll rip his eyes out ... until he's dead.' This embarrassing leaked video is followed by a video of him apologising to the public of Toronto.

But it doesn't stop there. There are more clips of him denying any problem with addiction, seemingly oblivious to his behaviour and choices. A clip of him insisting he's a normal, regular person, even as he's swearing in front of young schoolkids and making wildly inappropriate comments directly to the press, seemingly oblivious to what he is doing. Soundbite after soundbite setting the media alight, along with all the late-night comedy talk shows in America.

What was going on here? In a relatively short period, Ford's behaviour and choices seemed to take a turn for the worse. Mayor Ford died in 2016, two years after his term in office ended. His string of denials, poor judgements and rash decisions are the telltale signs of someone with drug, alcohol and/or behavioural addictions.

Psychologists and neuroscientists have run many experiments examining the effects of addiction on decision-making. Two main types of decision experiments have been run. The first is called the delayed discounting task and involves the kind of decisions we all make almost every day, such as between eating healthy food now, which is better for you long-term, and eating super-yummy food now that's worse in the long term.

The actual experiment asks questions like: Would you rather have twenty dollars today or fifty dollars in six months' time? More money is offered for the longer time-frame to compensate for waiting: if you get the money now you can spend it and enjoy what you buy for a longer period. Plus, who knows what will happen in six months? What we see in those with addictions is that they have a much stronger preference for choosing to take the money right now. In other words, a clear bias for impulsive actions with immediate rewards. Getting people with addictions to take the long-term option requires offering much more money.

The other experiment that has been used to investigate the effects of addiction on decision-making involves a gambling task. Participants are asked to pick a card from one of four decks of playing cards. With each pick, they bet on that card being a winner, but the chances of winning are controlled by a computer, and each deck has a different probability of winning. Some decks are safer and pay out more on average with small wins, while others are much riskier; they have a few huge wins, but also huge losses.

Individuals living with an addiction are more likely to choose the risky decks, and keep on choosing them, even after large losses, compared to those without an addiction. These results might seem obvious, but they needed to be demonstrated in a controlled lab setting. The experiments show that those with an addiction don't learn from bad outcomes – large losses – in the same way as the non-addicted. They seem not to be scared off to the same degree by the losses.

Interestingly, not only do we see these differences in decision-making, but there is also evidence that addiction is associated with differences in brain structure and function. A further twist to decision-making and addiction is that, presently, it's not entirely clear how much of the behaviour or brain differences is pre-existing and predisposes people to addiction. Or is it the other way around: does addiction cause these changes in behaviour and the brain? Or perhaps it's a mix of the two.

People with an addiction also have trouble linking their embodied emotional responses – interoception – with their conscious emotions. There is still much research to be done in this area, and neuroscientists haven't directly tested intuition in those with addiction. But from all the data we do have, it's safe to say that decision-making, in general, does not function normally during addiction.

If you're in an addictive relationship with a substance or a behaviour, then refrain from using your intuition; it most likely will not function well. Your decisions will be more impulsive and less connected with your emotions in a pro-ductive way. You are likely to have stronger impulses for things right now and to shrug off negative consequences that come later.

But when does a habit become an addiction? Are we all addicted to our social media? To coffee? It's outside the capacity of this book to go into detail about diagnosing an addiction, but generally speaking, clinical definitions of addiction often include terms like 'chronic, relapsing disorder', with disruption to social, psychological and physical wellbeing. You should seek professional help if you're worried about possible addiction.

To recap, instincts and intuition are two different things. The former are hard-wired and hence can sometimes be maladaptive, like fear of uncertainty. Cravings and addiction can also feel very much like intuition, but they are not. And if you're living with addiction, please don't attempt to use and develop your intuition.

Low Probability

SMILE

Resist the Temptation of Intuition
for Probabilistic Judgements

A game show that makes you angry

Humans are bad at probabilities. Our brains don't keep count of numbers the way a computer does. We don't experience them like we experience the smell or taste of coffee. Here's an example to prove the point.

We're on the set of the 1970s American game show *Let's Make a Deal*. The music chimes in and the audience cheers and waves sparkly homemade signs. The audience is dressed as aliens, sailors, cheerleaders, clowns, leprechauns, and anything else you can imagine; it's like Halloween.

Monty Hall, the host, comes in and the cheers and screams grow louder. From the audience he picks out a lady dressed in a suit covered in bananas. She stands up, shaking with excitement, stuttering when she talks. An assistant rolls in a flat-topped cart with three boxes on it.

'One of these three boxes, labelled A, B and C,' says Monty, 'contains the keys to that new 1975 Lincoln Continental.' He points and a velvet curtain rises on the stage to reveal the shiny red car, valued at US$11,000 at the time (around US$55,000 in today's money). The car glistens under the lights and the crowd goes wild. 'The other two boxes are empty,' Monty says. 'If you choose the box containing the keys, you win the car.'

He points to the three boxes on the cart in front of the audience. The contestant gasps. 'Select one of these boxes,' says Monty.

The contestant nervously chooses box B, and Monty offers her a hundred dollars for the box. When she turns it down, he ups it to two hundred. The audience screams *no*.

'Now remember,' Monty says. 'The probability of your box containing the keys to the car is one in three, and the probability of your box being empty is two in three.'

The contestant rejects his subsequent offer of five hundred for her box and he says, 'Okay, I'll do you a favour and open one of the remaining boxes on the table.' He opens box A. It's empty. The cheers and screams grow wilder still.

'Now,' Monty goes on. 'Either box C or your box B contains the car keys. I'll give you a thousand dollars cash for your box.' The audience lets out wild screams of 'No, no!'

'No thanks,' says the banana-dressed contestant, really getting into the mood now.

'Okay,' says Monty. 'One last chance: do you want to swap your box B for C?'

This time the audience goes quiet as they mutter and think to themselves. The contestant nervously sticks with her first choice; she and Monty open the box together, and she stares at the empty space inside. No keys. She puts her hands up to her head and bends over, as if she's been punched in the stomach. She then jumps up and down with frustration. She made the wrong choice.

What would you have done in that situation?

Let's go over the problem. There are three boxes. The car keys are in one. You don't know which one it is but the host, Monty, does know. And once a selection has been made he will never lift the box with the keys in it, as that would ruin the game. The question is, should you stick with your original choice or change?

Picture the two boxes, sitting there in front of you. If you are like me and the unlucky contestant (and indeed, most people), you will stick with the box you originally chose. There are two boxes left and therefore it's a fifty-fifty chance, right? Wrong. Statistically, you should switch boxes to win this game and switch every time.

For almost every non-statistician, understanding the probabilities in this game is difficult and frustrating, so let me unpack them. It illustrates just how bad we are at understanding probabilities. It won't hurt, I promise.

At the start of the game there are three boxes, and you have a one-in-three chance of picking the keys. Once you've made your choice, there's a two-in-three probability that the keys are in one of the other two boxes. When the host opens one of the boxes you did *not* pick, which will always be empty, the probability that originally covered both of the boxes you didn't pick is now concentrated on the one remaining box. That box now has a two-in-three probability of being the right one – twice as likely, in other words, to have the car keys in it as the box you originally chose. So, you should definitely switch any time you play the game.

This problem is now famous and is known as the Monty Hall problem. There are hundreds of explanations of this online – videos, blogs and statistical discussions attempting to get to the bottom of it. The problem has stumped people over and over since the 1970s. But why does the answer feel so wrong?

When we are faced with making quick decisions about probabilities like this one, we typically *feel* an answer. Some would say we intuit an answer, but it's not intuition. Since our brains are not equipped to deal with probabilities like this, we tend to get them wrong. In other words, it's another case of misintuition. This is why the L in SMILE is for low probabilities – we should never use our intuition when it comes to probabilistic thinking, particularly low probabilities. Think climate change or the risks of smoking for your health: understanding the probabilities in these issues can be difficult without carefully going over the numbers. Our brains, as I've said, just don't handle numbers very well.

Still not convinced? Here's another example. How many people do you think you would need to invite to a party and have actually turn up in order to have a 50 per cent chance of two of them having a birthday on the same day? Given that there are 365 days in the year, what number pops into your mind? Three hundred? A thousand? I bet it wasn't twenty-three, but that's in fact the right answer.

That *can't* be right, I hear you mutter under your breath. It's indeed the case that there are only twenty-two possible combinations of *your* birthday with someone else's, but that's not the important comparison here; what matters is comparing each person's birthday with every other person's birthday. You need to compare each person to each other person who is not you. The first person has twenty-two comparisons with others, then the second person has one less, twenty-one, because they have already been compared to the first person. The third person has twenty comparisons, and on it goes until you've added up all these numbers. This gives us a total of two hundred and fifty-three combinations. We simply can't picture this in our mind's eye, so we fail at understanding the probabilities and just think of comparing ourselves to the twenty-two others in the group.

As I said, we humans are just plain bad at probabilities. We tend to fall back on heuristics – mental shortcuts or rules of thumb that help us make quick decisions – feelings, or instincts based on experience and emotion. Psychology is full of the many ways we fail to understand probabilistic thinking. If we were to sit down and go over a probability with a pen and paper, we would likely understand it eventually. But we can't do it on the fly, not rapidly and not when we need it the most.

Heuristics often get confused with intuition, but they are not. Heuristics are simplified strategies that lessen the burden of deciding in the moment. They do not guarantee a correct solution, but they do significantly reduce the time and cognitive load needed to arrive at a decision. A recent example of a heuristic I noticed was at a café in which all the floor-length windows were slid open for summer, so you could easily walk straight in and out. However, everyone was still opening and closing the door to walk through, despite being able to take two steps around it and walk straight in. The heuristic here is that doors are how you enter, even if there's an easier, faster way available.

Another theory of cognition that relates to decision-making and gets associated with intuition is known as the 'System 1' and 'System 2' theory. This theory proposes that we have two main ways of making decisions: system one operates quickly and with little conscious effort, often relying on mental shortcuts, or heuristics. System two is deliberate, analytical, a logical mode of thinking about a decision. Psychologist Daniel Kahneman popularised the theory in his 2011 book *Thinking, Fast and Slow*, and in it he lumps everything to do with making decisions without conscious thought into system one, which people have associated with intuition. That is to say, it groups many very different cognitive processes (priming, cognitive biases, visual illusions, associative learning, intuition, and others) into one basket, which is confusing and not very scientific.

Many cognitive biases, for example, will induce rapid decisions that lead us astray, like needlessly opening and walking through the door in the above example. Our poor understanding of probabilities also tends to lead to bad judgements if we don't go over the numbers carefully and consciously. Our instincts will induce rapid decisions without much thought. Addiction and emotional thinking will also produce rapid decisions without deliberate analytical cognition – system one lumps all these different processes together. But they all have different characteristics and mechanisms, and are very different from productive and useful intuition as defined in this book.

Labelling all these very different processes as system one – or as intuition – is dangerous for decision-making, because it doesn't distinguish between processes that aid decision-making and those that hinder it; in other words, when it's safe to use intuition and when it's not.

Historically many in psychology have argued about whether intuition is a good or a bad thing when it comes to decision-making; this is in part because people clump so many different brain processes under the one banner and things get confused. The field of intuition needs a more detailed, nuanced theory, which separates out all these different brain processes and leads to more specific advice for decision-making.

Your fear of sharks isn't intuition

There's a famous stand-up routine by Jerry Seinfeld in which he addresses people's biggest fears. The thing that people reportedly fear the most is not death, but public speaking. This means, as Seinfeld says, that most people at a funeral would rather be the one in the casket than the one doing the eulogy.

It's not only funny, it speaks to a truth: we don't worry about or fear the things that are most likely to hurt or kill us, we fear other things instead. Even when we know the probabilities of danger, our fears still don't align with them.

Imagine you're floating effortlessly on salty water. Then the beach comes back into view, with people lying on the sand, splashing around in the shallows. Faintly, you hear something. It sounds like shouting. You flick your head around to see better. People are waving their arms and screaming. It's hard to make out what they're saying, the wind is carrying the sound away. Then you see a dark shadow move through the water under you. Was something there or was it just the shadows playing tricks? You frantically cram your chin down to your neck to get a better view, but can't make anything out.

You steal a quick glance back at the beach. People are really screaming now, jumping and waving. Shit, what are they saying? No, it can't be, but yes, it is, they are screaming, 'Shark!' You immediately turn and start swimming back in towards the beach.

Something hard hits your foot. It's rough like sandpaper. This time the dark shadow is unmistakable – the animal is there under you. You huff and puff, choking on the seawater as it splashes in your face. You smash your arms down into the water with a force you didn't know you had. You're in a kind of frenzy, mesmerised by self-preservation; nothing matters now but getting to the safety of the beach.

Then *whack*, like being hit by a car. The force is extreme, and you instinctively reach down with both hands, grasping to get at what holds you. Then you're under water, you tumble and spin like you're in a giant washing machine. Fighting, kicking, punching and clawing at whatever you can.

Apologies if that text got your heart racing. When we read things like this, we tend to conjure up vivid and graphic mental images of the event in our mind's eye, images that spike a strong fear reaction. We have run experiments like this in my laboratory in which people sit in a small dark room reading similar scary scenarios on a computer screen while we measure their body's physiological response. We've found that if someone has strong mental imagery (imagination) their emotional response to reading these scenarios is likewise very strong. However, if you have something called aphantasia – a blind mind's eye – then on average there really isn't much of a physiological response to reading scary stories.

If we have strong mental imagery, we create images in our mind's eye of scary graphic scenarios that seem to trick the other parts of our brain into thinking the scenario is actually happening. These graphic images induce activity in the brain's limbic system, which houses the circuits of fear and other emotions. What this means is that when something is easy to imagine, all else being equal, we are more likely to react strongly to it.

As a result, we fear things that are easy to imagine and not the things that probabilistically threaten us. We are also likely to worry more about such things than about those which are more likely to harm us. This is another strong example of how we don't understand probability well.

So next time you find yourself worrying about something graphic and easy to imagine, such as a plane crash or shark attack, remember that it's not your intuition talking to you. If you can, put that worry aside and look at the numbers. You are far more likely to die from being hit by lightning, or sun exposure, or the misuse of fireworks than from a shark.

What does this mean for intuition? If you have an intense fear of a plane crash, shark or terrorist attack, you shouldn't confuse these graphic, imagery-induced fears with intuition. This fear, like the bodily feelings of addiction, can be seductively similar to intuition. But it's not, and it will trick you into ignoring the low probability of such things actually happening. You need to avoid using your intuition to make decisions that involve things that are easy to imagine and therefore induce fear.

What *should* we worry about? If we take a probabilistic approach, then maybe we should look at the probabilities of death by different causes. The World Health Organization has data showing that cardiovascular disease is the global leader in causes of death. How many of us stress and worry about heart disease? Maybe some of us do, but not in the same way that we might worry about things like what other people think of us, work, spiders, snakes, or a great white shark looming out of the gloomy dark waters. Or yes, public speaking.

How to pick the winning horse every time

Here's an illustration of a different way in which the brain deals poorly with probability, this time from England.

Khadisha, a single mother working two jobs to make ends meet, gets an email one day asking if she wants to take part in a TV documentary about a failproof system for predicting the winners of horse races. Sounds like a scam, right? But the email also gives the first prediction from the mystery tipster, for a horse named Boz, and asks Khadisha not to place any bets but just to watch and see how this horse does in the upcoming race. It's a risk-free test of the system.

Khadisha later watches the race at home and Boz wins. Now she's starting to get interested.

Next, Khadisha receives a package containing a video camera (this is 2008, when phone cameras aren't yet that common), with which she's to record herself watching the races, placing bets, and receiving any winnings; this footage will be part of the documentary, she is told.

The second horse predicted to win is called Laced Up, and Khadisha receives the tip twenty-fours ahead of the race. Despite not being the favourite, Laced Up wins. This time, Khadisha places a bet, with her own money, and she wins twenty-eight pounds. Two from two so far – that's a success rate of 100 per cent. When the third tip arrives, Khadisha again places a bet, and watches the race live in the betting shop with the other punters. She's put twenty pounds on

Taunton Brook to win. The horse is an outsider at 18:1 but takes the win. Khadisha runs to the shop window to collect her winnings – three hundred and sixty pounds.

The fourth horse also wins, and when Khadisha is given the fifth prediction she goes to the racetrack to watch the race. Each time, the name of the horse was sent twenty-four hours before the race. The footage at the track shows her nervousness changing to disappointment, doubt, then fear as her horse is running last. Then, at the final jump, the leading horse trips on landing, causing the rest to also fall. All except Joe Lively, Khadisha's tip, which finishes out the race uncontested, the winner. The announcer screams, 'I can't believe it!' Khadisha can't either, but she now has five wins in row.

That same day, Khadisha is invited to meet the mystery tipster behind the system. In a private room at the race-course, sitting at a table, is Derren Brown, a famous British mentalist, illusionist and author, with many series of TV shows and stage performances under his belt. Khadisha recognises him immediately and is clearly shocked; she knows the kind of tricks he plays on people.

Nevertheless, she listens as he explains that his betting system can't fail. Then he tells her that when he gives her the horse for the sixth race he wants her to put a lot of money on it, because she won't lose.

'I want you to get together several thousand pounds,' he says. 'I will tell you how the system works, and you can choose if you want to continue using the system yourself; it's a lot of work so you may choose not to. But I will teach you how it works.'

Khadisha is persuaded. A significant win would be life-changing. But she doesn't have large sums of ready cash so she goes to her father, who's sceptical, but in the end decides to lend her a thousand pounds. As that's not enough, she then goes to a quick-cash loan company and borrows more. She puts together four thousand pounds.

Back at the racecourse once more, Derren confirms the tip: horse number two, Moon Over Miami. He offers to go and place her bet for her, saying he doesn't want her to see how much she stands to win. She pulls out the large wad of cash and nervously hands it over. Derren checks that she's definitely alright to do this, then leaves to put the bet on with a bookie.

When he returns with her potentially winning ticket, he explains his system. 'So a couple of months ago we reached out and made contact with you. Now, here's the thing, you weren't the only person we contacted. We actually contacted a very large group of people, just under eight thousand.'

Khadisha looks interested and a little worried.

'We started by randomly dividing 7,776 people into six equal groups. Six groups for six horses, one group for each horse in that first race.' All the people in the same group were given the same tip for the winner, but each group was given a different horse. Khadisha just happened to be in the winning group for that first race. All the people in the five losing groups then left the system. There were now 1,296 people in the system.

For the second race, the remaining people from the winning group were again split equally and randomly into six groups. Again, each group was given a random tip for one of six new horses in the next race, with each person in the same group getting the same tip. Again, the five losing groups left the show. And once again, purely by chance, Khadisha was in the winning group.

This process continued, with the losing groups dropping out, until there were only six people left for race five. Each of them was given a different horse.

As Derren explains this to Khadisha, the confusion, then disbelief is evident on her face. He then tells her that the previous week, at race five, she wasn't the only person being filmed for the show. 'Up until last week, the remaining six of you believed the system was foolproof.' Now Khadisha is the last remaining person in the system.

Race five, you'll recall, was the one in which every horse but Khadisha's fell over, but even if this hadn't happened, there still would have been just one winner, one person who had won five races in a row.

'It's simply a numbers game,' Derren says. 'You just happened, by chance, to be the person that won all five in a row.'

Khadisha's mouth opens as she realises there is not in fact a system for picking the winners, and she has just bet all her money and more on a random horse. Derren further explains there is actually no way of knowing which horse is going to win today.

'What's going through your head?' he asks her.

'Fucking hell, fucking hell,' says Khadisha, 'that's what's going through my head.'

The system was not a betting system, but a *belief system*. An elaborate set-up to convince people they can always win. Khadisha was so convinced she put up money she could not afford to risk, on a race that could not be predicted.

Low Probability

A person can undoubtedly win five horse races in a row, just by chance. There's a low probability of this happening, but it can happen. If it does, we typically, like Khadisha, don't think that it's just due to pure chance, we become convinced that something else is going on – that the system is not random at all, and that something magical might be happening.

'I feel sick, I feel faint,' says Khadisha as the race is about to start.

It doesn't go well from the get-go: Moon Over Miami, Khadisha's horse, is running at the back of the pack. And that's where he stays, coming in last. 'That's four grand gone,' says Khadisha. 'If the cameras weren't on, I'd be bloody crying now. I'm broke now, my dad is going to kill me.'

But there's a happy twist to the tale. Derren, with his typical sleight of hand, instead of placing Khadisha's money on Moon Over Miami, actually placed it on the race's winner. 'That's thirteen grand you've just won,' he tells her.

Khadisha won six races in a row against all the odds. When a rare event like this happens, we have trouble believing it's due to chance. Khadisha believed in a fictitious system that could always pick winners. This highlights another failure of the brain to understand probabilities.

When we think of random numbers, we rarely think of long strings of the same number, but these do occur naturally in random sequences; and given enough random numbers, they appear a lot. Interestingly, this is one of the methods used in fraud detection: fake data often omits long strings of repeated numbers, and this can be a giveaway.

So many people bet on horses daily that low-probability events like Khadisha's will happen. They happen a lot more when the overall number of events is huge. But the way our brains struggle with probability makes it very hard for us to accept this as random; it doesn't feel natural.

This is important when it comes to things like dream premonitions. Data suggests that, on average, we have about five dreams a night, or 1,825 dreams a year. If we remember only a tenth of our dreams, often when we think we don't dream it's just because they are quickly forgotten, then we recall 182.5 dreams a year. But strong emotional dreams about, say, plane crashes are more likely to be remembered.

There are around 68 million Brits, who have about 12 billion remembered dreams a year; the figure for Australia is about 5 billion dreams a year. The proportion of people who dream about plane crashes will be quite high, as it's a common phobia. This means that although the probability of an actual plane crash is very low – it's just over one for every hundred thousand commercial flight hours – on any given night there will be a lot of people dreaming about it. Hence, the probability of multiple people dreaming about a plane crash the night before it happens is low, but entirely possible.

Because we don't understand the probabilities, if we dream about a plane crash and one happens, it feels like there must be a better explanation than mere chance. Our brain searches for something else, and we may start to believe that we are special and have the power of premonition. This is called apophenia: the tendency to perceive a connection or meaningful patterns between unrelated or random things. The urge to believe that something special is happening is strong – it can feel like intuition but it's not. It's just plain probabilities.

Remember that the L in SMILE is to avoid making any decision that feels intuitive or automatic if it involves probabilistic thinking. I use 'Low probability' for the acronym, but the rule applies to *all* probabilistic thinking. Our brains don't learn or process probabilities like other things, and hence they can easily fool us. When it comes to probabilities, you're better off using your computer, phone or AI assistant.

Environment

SMIL**E**

Only Use Intuition in Familiar and Predictable Contexts

Usain Bolt in space

There's a gracefulness to the performance of top athletes, and the 100-metre sprinter is perhaps most graceful of all. Watching Usain Bolt dominate in that race, or the 200-metre sprint, is impressive. He glides with a strange kind of effortlessness, arms swiping by his sides like robotic chopping knives, passing the other runners in an instant to be way out in front by himself. Bolt stopped competing in 2017 but he still holds the world record for both distances. He is considered by many to be one of the greatest athletes of all time.

But there's one video of Bolt running that shows him fumbling at the start of a race. He stumbles, flailing his arms around, and has no advantage over the amateur runners either side of him. He bounces too high off the ground, then goes into a full somersault before crashing into a net. How can the greatest of all time perform no better than two amateurs?

This is a promotion video for Mumm Champagne. The point here is: context. Bolt is attempting to run in a low-gravity environment, on an aeroplane. The plane is flying in parabolas, moving through different levels of gravity, from 2g (two times the earth's gravity) to zero gravity. In the video, Bolt is trying to sprint the length of the plane with two others, and when the craft hits zero gravity, he flies up to hit the ceiling.

All Bolt's years of running and of being literally the best in the world don't help him run in this different environment. The weight of each of his limbs is different, so he moves his arms and legs too quickly, and there is less friction with the floor under his feet. He pushes off the ground too much, bouncing in the air the way you see in old moon-walk footage. Changing the environment has completely thrown off a lifetime of learning.

Put you or me in a similar low-gravity environment, and all the countless things we never think of while moving would now require careful attention and would have to be learnt all over again.

It's the same with intuition: when you change your environment, much of the learning linking things in the environment to possible outcomes must be relearnt. Whether it be physical movements or decision-making, doing something in a different context can invalidate the prior learning behind it and hence invalidate your intuition. The hundreds of subtle clues that your brain has learnt over the years as to which things in the environment predict what are suddenly no longer relevant.

Changing the context or environment can change everything. Non-verbal cues provide powerful examples of this: few things, in fact, hold as much silent power. Imagine navigating a bustling street in New York, effortlessly throwing a thumbs-up to a street vendor who gets your hot dog order just right. This very gesture, so benign and filled with positive intent in most Western cultures, could be your downfall in the streets of Baghdad. Here, it's not a salute of approval but a blatant expression of disrespect, akin to the Western middle finger. Its very meaning has altered – and not in your favour.

Now picture yourself in a Bulgarian boardroom, nodding to show agreement in a crucial business deal. To your shock, the deal falls through. Why? In Bulgaria, nodding doesn't mean 'yes', but quite the opposite: it means 'no'. A single head movement, so instinctual and automatic to many of us, has a completely different consequence in a different environment.

The okay sign, made by connecting the thumb and index finger, is another that in a different environment has very different meanings. This sign screams 'worthless' in France, and is considered obscenely offensive in Brazil and Germany, and elsewhere. Imagine flashing this sign in a Brazilian bar after enjoying a fantastic samba performance; you might just find yourself facing a confrontational dancer – not the kind of encore you'd have hoped for.

So, the next time you find yourself in unfamiliar terrain, remember: the environment is not just a silent spectator. It's an active player, reshaping, redefining and recontextualising the learnings behind intuition. Navigate wisely.

Many of us have good intuitive navigation skills, whether in the city or out in nature. Take, for instance, the fact that moss, which requires cool, wet conditions, grows predominantly on the southern side of trees in Australia because the sun's arc leaves the southern side in the shade. This is something that can be used intuitively for navigation. But transport someone to the forests of the Northern Hemisphere, and suddenly their intuitive compass might be all off. There, moss prefers a tree's northern side.

There are hundreds of other subtle changes in the natural environment between hemispheres, and even if you're informed of them your intuition may not update itself, leaving you vulnerable to misintuition and getting lost. These changes in environmental cues serve as humbling reminders that our intuitive skills are not universal constants, but localised adaptations.

Why you should study under water

In 1975, near the city of Oban in Scotland, famous for its whisky, two divers are preparing to drop into the water. They are part of a now-famous experiment involving learning under water. The day is cold. The wind blows strongly off the dark water, the sky ripples with occasional white light piercing the thin patches of cloud. The divers sit on the wharf in wetsuits and diving gear: scuba tanks, masks, and a strange-looking set of headphones, nestled between their masks and the mask straps. Instead of sitting over their ears, these headphones sit just in front of the ears, towards the face. They are bone conductance headphones; they transmit sound not through the air to the eardrums, but via the bone of the skull. This enables the wearer to hear while fully submerged under water.

The divers have heavy weights strapped around their waists, and once in the water they sink quickly to the bottom. Sitting comfortably on the sand, they retrieve small whiteboards with pencils attached. Through their headphones, someone at the surface asks if they're ready to begin the experiment. They give the okay sign to a third diver already in the water, and a prerecording starts to play.

First, they're asked to breathe in a particular rhythm, to control the underwater noise. Breathing with air regulators under water is very noisy, and these divers need silence between breaths so they can hear the words they're to be tested on. Once they have the right breathing pattern, they hear three unrelated words, then the word 'breathe', at which they take a nice slow breath, followed by three more unrelated words and a chance to breathe again, then three more words.

Together the divers were later tested, either under water again or on dry land close by, to see how well they remembered the words. As part of the experiment, other groups of divers were first given the words on dry land and then later tested on land or under water, eighteen people in total.

The study found something interesting. Those who learnt the words under water remembered more of them when they were back under water again, compared to when they were tested on the surface. Whereas those who learnt on dry land remembered more words when tested on land and fewer when they were sitting on the bottom of the ocean. In other words, where you learn matters. If you learn and remember in the same place, you remember more, even if it's in a strange place like the bottom of the ocean.

This effect is called context-dependent memory. The context in which you learn something is attached to that thing in your memory, in such a way that being back in the same place helps the memory come back. You may have noticed, for example, that when sitting in an exam room, it always seems harder to remember the things you learnt at home. All that juicy knowledge you crammed in right before the exam also included your bedroom or office along with it. The knowledge was that little bit harder to access in a different place.

Almost everything in your home-study environment can matter: the music you might be playing, smells, how you were dressed, the layout of your room. That's why exam-prep advice will often include tips like using a specific oil or fragrance when cramming at home. Applying that same smell to yourself during the exam will make that environment more similar to your home's, and hence boost memory recall. In fact, there's good evidence that simply chewing gum while you learn is enough to create significant context-dependent memory. But you must make sure that you chew gum during both learning and your exam, or not at all.

This is the reason that environment, or context, matters for intuition. The hundreds of hours of learning in one location will not generalise well to different locations. Changing the context will induce possible mismatches between predictors in the environment and their positive or negative outcomes – the engine behind intuition. And even if those predictors are still working, their signals will be weaker in a different context, and again your intuition will be affected.

Importantly, the environment in which something is learnt includes the state of your body and brain. Research has shown that context-dependent memory also involves what is called state-dependent memory. The state you are in when you learn something changes how information is encoded in your brain, making it easier to remember something when you're in that same state. All those jokes about getting drunk again to remember where you left your phone or keys the night before are actually based on science. To a degree, you can indeed better remember things you learnt when drunk if you're drunk again, but it's worth noting that these effects will depend to some degree on what you are learning and how intoxicated you are.

The mood you're in also affects state-dependent memory. Whether you're tired, highly caffeinated, sad, in pain, or in any other state, you will have better memory retrieval and use of learnt information when you're in that same state once more. Different internal states are just like different external places when it comes to learning. That's why jet fighter pilots like Jason from the introduction need to practise under stressful scenarios that simulate conflict and danger instead of in more relaxed conditions. It ensures that state-dependent memory will be maximised.

It's because intuition is based on *all* the unconscious associations your brain has learnt between cues and what they predict that it's so important to rely on your intuition only when you're in a familiar context, externally and internally.

Intuition in an uncertain world

Many environments are predictable, and hence there's plenty for you and your intuition to learn. But in those environments that are unpredictable, when things occur randomly, your brain can't learn any patterns and intuition therefore can't work. But that won't stop your brain from trying.

A great example of an unpredictable environment is a casino. You lean into the roulette table, hands resting on the shiny mahogany edge. You've just won some money by betting on black. You bet half your winnings again on black; the wheel spins – bounce, bounce – and the winner is black again, so you win again. What the hell, one more time on black, same again. You win again. That's three blacks in a row, and three wins in a row. You wonder: 'Should I go again?' One more bet on lucky black can't hurt, you lay down the chips, black again, and oh my god, you win again.

Four blacks in a row, what are the odds? Surely that's it, the streak has to finish now. What to do? That nagging voice in your head pushes you to go one more time. Just do it. So, down the chips go, on black one more time. You hold your breath, the ball bounces around – black again. You let out a huge sigh. Okay, five blacks in a row. This crazy trend just can't continue. So you, like all the others at the table now, flip strategies and go with red. The crowd gasps: black it is again, you lose.

That's now six blacks in a row. What to do now? It's surely defying physics. So again, you and the crowd bet on red. But alas, black wins again. Seven blacks in a row. The sounds of astonishment from around the table have drawn a larger crowd. The piles of chips on red are getting larger and larger. Eight, nine, ten blacks in a row. People are laughing and crying, not able to believe what they're seeing. With each new black spin, the crowd's confidence builds that the next spin must be, has to be, red.

Eleven, twelve, thirteen, fourteen, and yes, fifteen blacks, without any sign of red. This has to be a world record. People are talking; is the table broken? The bets on red keep getting larger and larger. People just can't believe what they're seeing. Sixteen, seventeen, eighteen, nineteen, twenty blacks in a row. Absolutely convinced that a red is now due, you slide over everything you have onto red. But, no, yet another black, and you've lost it all.

Twenty-two, twenty-three, twenty-four. You've lost all your chips, so you can't bet anymore, but others are still hard at it; the piles of chips on red are now huge, in the millions of dollars. Twenty-five, twenty-six blacks, the crowd screams, the attendant laughs and shrugs, and his supervisor stands resolute. Nothing like this has ever happened before. Millions and millions of dollars have flowed into the casino's hands because people were convinced a red was due on each new spin.

Then it finally happens, on the twenty-seventh spin, bounce, bounce, red. The streak is over, but it's left most of the crowd penniless.

On 18 August 1913, this actually happened at a Monte Carlo roulette table. The ball fell on black twenty-six times in a row, an extremely uncommon occurrence. Gamblers lost millions betting against black, reasoning incorrectly that the streak was due to end. This is known as the Monte Carlo fallacy or gambler's fallacy. The belief that each new spin of the wheel is somehow related to the previous spins, so that when you get a string of blacks it somehow means a red is due to balance out the odds.

The critical point here is that for unpredictable random environments, we should not pay attention to any feelings we have that might resemble intuition, because our brains can't learn any patterns for what variables predict which outcomes. But here's the catch: even when we know that the environment is not predictable, we still often feel like it is.

These kinds of false beliefs are up there with super-stitions. We can't help ourselves, we try to predict patterns in randomness. That little feeling builds up inside you that the next spin will be a winner because of the past few. Or maybe because you crossed your fingers, uncrossed your legs, and said a little prayer to yourself. All these types of gambler's fallacies are just that, fallacies. They are not your intuition picking up on the subtle trends of the table, con-necting to the blueprint of the universe. Intuition can't work with random events or in uncertain systems. As I men-tioned earlier, the cognitive bias to see patterns in things where there are none is called apophenia, and it's very hard to turn off.

Usain Bolt can do amazing things running on earth, in earth's gravity, and he could probably learn to run in the moon's gravity. However, if the strength of gravity were to keep changing each day, he wouldn't be able to learn to run fast because the required learnings would also constantly change. It's the same for intuition. In unstable and unpredictable environments, we need to be very careful about using intuition, because it's equivalent to continually changing the environment.

Many indicators suggest that the world is becoming less and less predictable, both in terms of positive, exciting technological advances, and also extreme disasters – so-called Black Swan events, pandemics, wars and climate events. This can be very uncomfortable for many people, as neuroscience experiments have shown that the brains of humans (and animals) respond to uncertainty as a source of fear – a bit like looking over a cliff or seeing a dangerous snake or spider. Uncertainty is literally a fear stimulus, it drives fear in the brain – in some more than others. If our modern environment is becoming less predictable, what does this mean for using our intuition?

It is important here to define the types of uncertainty and consider how they relate to intuition. Let's talk a little about business dynamics. New technology is causing exponential change, from autonomous AI and crypto-based transactions to longevity and advances in biology. We are seeing the kind of change that shoots up on a graph like the end of a hockey stick, as the adoption of these new developments increases exponentially over time. What's more, each one of these individual exponential curves is beginning to interact in ways almost impossible to predict. Does this mean intuition will fail?

No, on the contrary. With limited data and the need for more rapid decisions, intuition has never been more important to business. The modern environment, although more uncertain, is not random. If you've been working for some time developing new products and services you'll have seen, like Steve Jobs, how quickly new products can take off or fail. Your brain has probably also learnt which characteristics predict positive or negative outcomes. The unconscious associations behind intuition should already be there from your mastery. Hence, as the world becomes less certain, there's no reason why intuition won't still work – the world is not becoming random, just harder to predict.

In contrast, we don't have much experience with global pandemics and large-scale global health challenges or climate change, so our brains probably haven't learnt the underlying associations to drive intuition in these areas. So, for unfamiliar sources of uncertainty, we should stay away from our intuition, at least at first. As we are exposed to more climate events, we might develop better intuition for how to respond to storms, floods and heatwaves. But until then, we need to analyse the scientific data and make rational, logical predictions based on that data.

Bias in intuition

Another aspect relevant to the context of learning involves watching television or films and reading. While absorbing content in this way is not going to induce the same powerful learning as actually living through the events yourself would, if you spend enough hours binge-streaming or reading, you will induce some learning, and that learning *will* affect intuition.

Think of the amount of time you've spent this way in the past few years. Maybe you watched all twenty-five James

Bond films? Maybe the full Marvel cinematic universe (more than once)? *The Lord of the Rings*? Were there subtle or perhaps not-so-subtle biases in this content? Gender biases? Race biases?

Even if the content you've watched is fictional, the biases in its content can have subtle effects on you. Consume enough biased content and this will influence the associations that make up your intuition.

The accusations of gender and political bias in AI are a good example here. Just like us, AI has to first learn things in order to make use of the knowledge. It must sift through masses of data and learn by itself, a process known as unsupervised learning. Any biases in the datasets AI is given to learn from will be inherited. For example, multiple recruitment firms, and companies themselves, have trained AIs in an attempt to better select high performers for job success, and such training often included variables like personalities, cognitive tests and all kinds of rich metrics. When the AIs were let loose on applicants it was apparent that mostly males were being selected for the roles. That was because the data the AIs were given had more males in high-ranking roles than females. The early versions of ChatGPT were also trained on datasets from the internet, and so inherited the biases on the internet.

It's the same for your intuition. If you train your intuition on biased data, whether from real life or fiction, your intuition will inherit the bias. Our hidden race, gender or age biases are the potential dark side of intuition. If your intuition is trained from a workforce with more males and older people, then it should be no surprise that it replicates these biases.

The rapidly emerging trend of synthetic media – digitally generated or manipulated multimedia content such as images, audio, video or text, including those created by AI – will almost certainly bring with it intentional and unintentional biases of all kinds.

So, when it comes to the E in SMILE, you need to be aware of the potential for bias in the environment in which your intuition was learnt. Training in a biased environment will perpetuate the bias and won't nudge you towards optimal decisions. You might hire the wrong person for a job, or under- or overestimate someone's abilities.

It's also the case that who we choose to spend time with will affect our intuition. If you spend a lot of time with people with a particular bias, your intuition will likely pick it up and become similarly biased.

PART THREE

The Practice of Intuition

A daily practice

I use intuition many times a day, every day. I am not embarrassed to admit this because there's nothing woo-woo about intuition. Indeed, I use it for my work as a scientist, which some people seem to find ironic. There is a prevailing idea that because science is rational and follows rules, you can't use intuition for science – but this couldn't be further from the truth. Intuition, as this book has shown, is utilising unconscious information to make better decisions and actions. Who wouldn't want to have more information at their fingertips if that information is genuinely helpful? It's simply an advantage.

Whether reading a new scientific paper that's just been published or marking a student's paper, my intuition is hard at work. Often when reading a paper, I'll get a nagging feeling in my body that something just isn't right with the experiment, things don't line up, something is off. I'll take a pause, note that feeling, and continue reading. My intuition is picking up on signs in the text, based on the hundreds of prior papers I've read over many years that have predicted good or bad science, poor definitions, and confounded or problematic experiments. There are so many cues in the written text that my brain is processing that I don't try to rationally go over them, or figure out what exactly they are. I feel the feelings, note them, continue reading, and see how predictive they are. But I don't use intuition when it comes to mathematical models, statistics or probabilities.

I also use it for my own science. In the worlds of psychology and neuroscience, there are almost infinite options in running new experiments. How do I choose where to invest the lab's time and energy? Intuition. Importantly, using intuition doesn't exclude also using conscious rational logic; the two can work well together. I also use intuition to help decide who I work with, and yes, I used intuition in deciding to write this book.

However, it's also worth mentioning that there are many times when I avoid using my intuition. When SMILE is not met, I actively avoid and try to ignore feelings of intuition. When I'm emotional, stressed or anxious, I will put aside any intuitive feelings and look at the logical pros and cons of a decision. This can sometimes be very difficult – something I imagine to be like flying by instruments, to use pilots' terminology. When your emotions are kicking up, you feel anxious, and that voice is saying, 'Is this really the right thing to do?' The fear kicks in and mixes with the anxiety, and all you feel is: 'Don't do it, something is wrong with that option.' Here, my friends, is where we have to turn to science. We must trust the numbers and not our biased emotions, our anxious emotional thinking – that's not intuition.

When I'm dealing with probabilities, both at work and in my private life, I also try to avoid using my intuition. I must admit that one of the rules I find hardest to follow is the E for environment. I find it difficult to remember not to use my intuition in different and novel settings. When travelling, I take care to not blindly follow my social intuition, or geography, or feelings about the weather. The habit of automatically using intuition can be hard to break, hence the importance of the five rules of intuition.

When it comes to beginning a practice of intuition, it's crucial to remember that it's not about jumping into the deep end. Don't start with huge, life-changing decisions – even though it's often hard to avoid feelings of intuition in those scenarios. You need to build a daily practice that will help you train your intuition and keep track of your progress, and to do so slowly and safely.

Let me take you through some examples of everyday decisions and how you could apply the five rules of SMILE to each.

You enter a cosy bookstore, a place you've been before. The scent of fresh paper and ink envelops you, whispering tales of adventure and wisdom. Today, your mission is to buy a new book to read, but you don't have one in mind, nor do you have any particular genre or topic in mind.

Walking through the aisles, you feel emotionally grounded. Recent events and your hard work have left you in a place of equilibrium, neither overly excited nor deeply pensive. You're reminded of the first rule of intuition, self-awareness: with this box ticked, you walk on.

Your history with books, having read hundreds or thousands, has equipped you with a keen sense when it comes to discerning one title from another. The covers speak to you in a language only true book lovers can comprehend. Here, the second rule becomes evident, mastery: your intuition shines brightest in fields where you've amassed knowledge and experience.

A contemporary section introduces books integrated with a social media platform. Each purchase guarantees exclusive access to vibrant online communities, with discussions and real-time reviews. The thrill of notifications and the charm of being part of a buzzing community to show off to tempts, but the third rule in SMILE echoes in your mind: be wary of trusting intuition in areas engineered for addictive allure, like social media. So you walk on.

A banner promoting a new membership programme catches your eye as you wander further. *Sign up and stand a 30% chance of securing a 90% discount on your next buy, or a 50% probability of a 25% reduction!* (Okay, I've exaggerated the banner a little, to make the point.) While enticing, you remember the fourth rule: intuition isn't the best tool for making decisions based on probabilities, so you walk on past.

INTUITION

The ambience of the bookstore, plus memories of other libraries and reading nooks you've frequented, assures you. The rustle of pages turning, muted conversations about favourite authors, the familiar surroundings, all make you feel at home. The fifth rule is clear in this moment: your intuition is most reliable in environments you know well, and you know bookshops and libraries well.

Guided by these principles, you're drawn to a book with a subtle yet captivating cover. The blurb resonates, and something nudges you to buy it. Holding the potential treasure, you head to the counter, eager to dive into its pages.

Here's another everyday example. Time for a holiday? You find yourself longing to reconnect with nature. Digital screens and browser tabs present you with a plethora of options: tranquil lakeside cabins, majestic mountain trails, sunny beaches. The excess of choices could be overwhelming, but today, you've decided to let intuition guide you.

As you peruse your options, you take a moment to centre yourself, recalling the serene moments from past getaways. The weight of the workweek seems distant, allowing you to be in the present. You check in and assess your emotional state – you feel calm and balanced. With this emotional grounding, you trust your inner compass.

The Practice of Intuition

Years of travelling, whether it be short road trips or extensive international travel, have imbued you with a seasoned traveller's insight. You've booked from websites, with travel agents, and on the advice of friends. You have plenty of expertise in searching, choosing and predicting the types of trips you like. This is where the second rule shines: you have mastery in travel booking from years of experience.

However, just as you're about to finalise your plans, a niggling worry grips you: 'Can I truly afford to take this time off?' The uncertainty of the future of your work, the looming shadow of AI, and the changing dynamics of the employment landscape fill you with uncertainty. But then the third rule comes to mind, grounding you: not all internal sensations are intuitive. That fearful response, ingrained through millennia of evolutionary pressures, is an innate reaction to uncertainty. It's not intuition, but a relic from a time when every unknown might have been a potential threat. Recognising it as a maladaptive response in this modern context, you set it aside.

Just as you're nearly set on a forest retreat, an enticing ad for a diving holiday pulls you in. The allure of vibrant coral reefs and the mysteries of the deep beckons. But as you imagine yourself submerged, a brief pang of fear about sharks surfaces. Images from movies and sensational-ised news stories come to mind. Then you remember the fourth rule, about probabilities, and how we so often get them wrong: decisions should be based on actual odds, not imagined fearful scenarios. A quick shake of the head reminds you that shark attacks are incredibly rare, helping you set aside the apprehensive feeling.

While you have vast experience with booking and trav-elling, most of that has been for work trips, so you remind yourself of the final rule, about the environment or context of your intuitive decisions. This is a personal holiday you're planning, and you realise that any intuition you might have developed for making travel decisions might not transfer from work to holidays. So, you logically and rationally go through each option before making a decision, relying less on your intuition.

On my website (www.profjoelpearson.com) you can find a digital version of the chart shown on the following page. I suggest either printing it out and putting it in a place where you'll see it often, such as on your fridge or above your computer, or putting a digital version of it on your phone.

To use the chart, start at the left with three deep breaths, then follow the curvy SMILE down and across to the right.

Self-awareness
Mastery
Impulses and Addiction
Low probability
Environment

SMILE

Practise intuition

E
Is this a new environment?

No

Yes

L
Does this require probabilistic thinking?

No

Yes

Don't practise intuition

I
Is this an impulse or craving?

Yes

No

M
Still in the learning phase?

Yes

No

S
Are your emotions too high?

Yes

No

Take three deep breaths

Think of this as a checklist: if you get a yes at any point, then stop and hold off on practising intuition. Come back to it when your situation has changed.

The idea is to have the diagram close at hand and go through each step of SMILE every time you're going to use intuition – at least at first. After some time you'll notice that the checklist becomes embedded in your memory and you won't need the diagram as much. You'll be able to go through each letter yourself mentally. Then, in time, you won't need to check each letter individually, the process will become automatic. You will just get a sense of whether SMILE is met or not.

Think about this process as being like going to the gym. At first, someone has to teach you how to use the machines and weights, how many reps and sets to do and when to do them. You follow a printout or app for the first few weeks, ticking off each set as you go. Over time, you no longer need the sheet or app because you remember your routine: how much weight you were lifting, how many reps you were doing, how to do the exercises. This progression also applies to developing mastery of SMILE and safe intuition.

To optimise your intuition practice, I recommend keeping track of your intuition: what you used it for, how it felt, and whether it was successful or not. People often think that human memory is like a video recording of what happened each day – but that's not how memory works. Our memory is biased in so many ways. Basically, our brain constructs a new version of events from the bits of information that it's stored. We tend to remember the emotional high points and are biased for items at the beginning and end of things. The more we replay our memories, the less reliable they become. By reliving our memories in our imagination, we actually alter them bit by bit without realising it.

All this is to say that if you want to keep an accurate record of how your intuition is performing, then you need to take notes to record it – don't rely solely on your memory. The old saying, what we measure gets improved, rings true here.

Keeping track of your intuition in an intuition journal or app is a good idea. On the following page is a simple example of a table you can use to track some of the important parts of the intuition process. It shows eight separate stages, and uses the example of deciding which café to go to.

DECISION	LOCATION OF SENSATION	STRENGTH OF SENSATION 1–10	CHOICE	OUTCOME	STRENGTH OF OUTCOME 1–10	SUCCESS	CONTEXT
Café A or Café B	Gut and chest	6	Café B	Enjoyed, happy, surprised	8	yes	Chose Café B at home, in a calm state of mind

- **Decision:** Note down what the decision is between (what the options are). In the given example, it's between going to café A vs going to café B. There might be only two options, or there might be many. Write down the options you know about or that pass through your mind.

- **Location of sensation:** Note where in the body you feel any intuition. In the example, it's in the gut and chest, but it might be a tingling in the hands or somewhere in the back your head. There doesn't have to be just one location, the sensation could be in many different spots throughout your body. And if you can't pinpoint any particular location, note that down too: that the location was ambiguous, or the feeling had no bodily location.

- **Strength of sensation:** Rate how strongly you feel that sensation, from 1–10. With one being something you were only just aware of, something very weak and subtle, and ten being the strongest sensation such as a strong sinking feeling in the gut like Jon experienced on Everest. Or maybe you feel nausea, an itchy discomfort, a strong sensation that something just isn't right.

- **Choice:** Note down the choice you made; in this case, café B. You don't need much description here, just the choice or action you ended up making.
- **Outcome:** Note how you felt about the choice you made. Were you glad you made it? Was it a good or a bad outcome? Were you surprised by the outcome? Think about this cell as the feeling that drives the learning behind your intuition. Remember that one way to estimate whether you've reached mastery is when you have fewer and fewer surprises. This is a good cell to keep track of whether or not you were surprised with the outcome.
- **Strength of that outcome:** Record the strength of your feelings about the outcome. How happy or unhappy were you with your choice? How surprised were you? The numerical ratings in this cell refer back to the Outcomes recorded in the prior cell. If you noted down both happiness and surprise in the Outcome cell, you would rate both of those separately here.
- **Success:** A categorical assessment of the overall performance of your intuition. Did your intuition lead you in the right direction? In other words, was it

successful? Over time, the idea is to have more and more 'yes' answers in this column.

- **Context:** Note where you used your intuition, remembering that the learning behind intuition is context-specific, in respect to both the physical environment and the internal environment of your body. In this cell it could also be useful to note down any specific or unusual states you were in. For example, whether you were highly caffeinated, or had partaken of a few alcoholic drinks, this kind of thing.

When starting your intuition practice, it's important to be mindful of feedback loops. In the chapter on the second rule of SMILE, mastery, we saw how timing matters for learning intuition. Well, it's relevant to your daily practice as well. If you choose decisions that take a long time to give feedback, such as buying a house or long-term investments, you'll have to wait a long time to know if your decision was successful or not. This will make keeping track of your success hard. To begin with, it's better to choose decisions that have a tight feedback loop, decisions that give immediate feedback. This will enable you to make more decisions over a given period, have more iterations of the whole process and thus more chances to learn. Practising intuition while playing a sport you've mastered is a great example here; you may not have time to jot down each instance of intuition during a game, but you will get a lot of chances to practise it.

I like to run on bush trails, and on good days where SMILE lines up, I practise my intuition. I focus on my intuitive feelings for different foot placements on rocks and then try to pay attention to how my intuition predicts a stable foothold or a slip and a bruised knee. However, on days when I'm emotional or running in an unfamiliar environment, those intuitive footholds are more likely to slip. The idea is to practise linking a feeling associated with unconscious information to practical positive intuition outcomes, and cultivate those links to build them up.

The optimal times for intuition

While this book is about understanding what intuition is and when it's safe to use it, some everyday scenarios are naturally more suited to using intuition than others. There are no absolutes or strict rules here, but there are general trends for when intuition can be more useful.

As long as you've checked off all five SMILE rules, situations where time and information are limited are some of the optimal moments to use intuition – precisely because you don't have the time or data to use a conscious analytical strategy. When you're in a rush, when someone passes you the ball in a game, you simply don't have time to rationally cogitate over all the possible options. You need to act fast and rely on intuition. You must feel where to run to next or who to pass the ball to.

That is to say, time pressure is a big one here. Anytime you have to make a rapid choice, intuition is likely to beat conscious logical strategies. In situations where you have plenty of time to mull over the data, draw up pros and cons, and think through a strategy, you don't need to rely on intuition as much.

The other dimension that affects decision-making is the amount or type of information you have. When information is limited or ambiguous, it makes good sense to fire up your intuition and use it. Interestingly, intuition data from my lab using the Emotional Inception experiment discussed in Part One's chapter 'Measuring Intuition', shows that intuition can boost decision accuracy more when the conscious information is ambiguous.

This brings up an interesting point regarding modern decision-making in our uncertain environments. Many people in business talk about the need to make faster decisions with less information at hand. This, at least in theory, suggests that decision-making might be more suited to using intuition than ever before. Or to put it another way, our work environments require intuition more and more, due to the agile nature of modern business.

Artistic and creative decisions are also more suited to intuition. Imagine you're in an art gallery, standing in front of a canvas bursting with colour, emotion and expression. You are called on to judge the quality of the artwork – this is an artistic or creative decision. It's a holistic judgement that can't really be broken down into a checklist of components or sub-decisions. There is no good and useful way to tally the colours, brushstrokes, perspective or references to prior works. You have to feel it. You have to take it all in at once and make a call about the artwork in its entirety. This type of decision is non-decomposable, it can't be broken down into parts. It's another example of a good time to lean on your intuition.

Likewise, think of a fast-paced environment – a corporate boardroom, for instance. Here you face complex, multidimensional and interdependent problem-solving, where the issues are interwoven and have countless variables. These decisions aren't decomposable either; you can't make them in the way you'd code a software program or assemble a piece of furniture, where each step can be neatly broken down and executed. In the boardroom, you're playing multidimensional chess in real-time. Your intuition is often your best bet, offering possible clarity amidst the chaos, a way to navigate the murky waters of ambiguous information. Of course, it goes without saying that the SMILE boxes must all be checked.

Lastly, imagine a crisis – a house on fire, a company teetering on the brink of bankruptcy, or a tense political standoff. These are emergent situations where time and data are a luxury you can't afford, similar to sports. There's no time for decomposable tasks, like planning a project timeline or creating a budget. These types of scenarios are also well suited to intuition.

To summarise, once you've developed your intuition, it tends to be more useful for complex, holistic decisions that cannot easily be broken down into smaller sub-tasks.

It's also useful in situations where there is ambiguity, uncertainty, or a lack of complete information, and where logical or analytical thinking might not yield a clear solution. When these situations also have a limited timeframe, then intuition can be even more important.

Should you use intuition for making life-changing decisions like getting married, buying a house, or getting a divorce? For these hugely impactful decisions people often report falling back on intuitive feelings, despite having an overwhelming abundance of logical conscious information.

An interesting strategy for reversible decisions is to first make a non-binding decision, then see how you feel about it. In other words, you reverse the normal order for making intuition-based decisions. Do you now have a sinking feeling in the stomach? Do you feel a little off? If the decision is an irreversible one, you can try to imagine or act as if you have made it and see if this prompts an intuitive response.

There are a few key things to stress yet again here. First, ensure that all SMILE boxes are ticked. Second, if you're new to using intuition or unsure about your intuition, don't start with large, life-changing decisions. Start with the small stuff – think tea or coffee, this café or that one – and work your way up to the big stuff. People often report taking note of gut feelings during large decisions. This doesn't mean they've relied solely on intuition to make the decisions, but when the stakes are high, gut feelings can be strong and hard to ignore.

Importantly, we have to take extra care with S from SMILE – self-awareness – to ensure we are not in an emotional, stressed or anxious state when we make large decisions. For big, life-changing decisions with a lot on the line, anxiety can creep in because of the high stakes. If your emotions are fired up, then take a moment or two, and try to level out; this will help you get back in touch with your intuition.

Perhaps a good rule of thumb when it comes to large decisions is to try to blend both intuition and rational logical analysis of the data. Compare the outcomes you get using these two approaches and take note of where they align or differ. Are they in opposition to each other? Do you get a gut response to the logical, conscious conclusion? How does it feel? Might it prompt you to re-evaluate the logic?

Some people naturally use intuition more

Some individuals appear to be inherently gifted in their intuitive abilities, much like a musician who can play by ear, or a basketball player who can effortlessly sink a half-court shot. These people find themselves intuitively navigating life's challenges; others, well, not so much. In the data from the intuition experiments done in my lab, we found that some individuals could leverage unconscious emotional images to improve their conscious decision-making, while others were unaided by unconscious information.

Why is this? Why do some of us readily use intuition while others do not? Unfortunately, this is another thing we just don't yet know. Perhaps the answer is tied to our different sensitivities to our internal bodily sensations, interoception; some people are much more sensitive to interoception than others. Or perhaps it's because some people learn more quickly than others. Maybe it's got to do with how susceptible we are to emotional cues, or how well we process things perceptually.

But just because you don't currently use intuition, or feel that you can't use it, doesn't mean you can't learn to. Our data from the Emotional Inception experiments shows that intuition, much like any other skill, can be learnt and improved with practice – the more you do it, the better you get.

Across the range of different intuition abilities, who amongst us has better intuition? There's a lot of talk about mothers' intuition and female intuition, and many people want to know if there's evidence of that. There is evidence that women report using intuition in their decisions more than men do, while men report using more rational strategies for their decision-making. However, there is not much data on this from objective measurement techniques, like our lab-based intuition task.

In a related vein, there is data showing more so-called magical thinking (beliefs regarding the existence of phenomena such as ghosts) in women. These trends have been discussed in research papers in terms of the documented differences in emotional intensity, with women, on average, reporting their emotional experiences as more intense than men. However, the differences in socialisation have also been written about as a possible contributor to these differences. Men have often been socialised to be more rational and to disavow the use of emotions for decision-making. An interesting twist to this is that men frequently show higher levels of superstitious behaviours in relation to sports and gambling than do women.

There is some evidence to suggest that East Asian cultures rate intuitive reasoning as more important than analytical, rational decision-making, but this is another area of research that needs further investigation.

To sum up, despite a range of inherent intuition abilities, there is no reason why anyone can't start a practice of intuition. So let's not view this as a dead end, but rather as an invitation. An invitation to hone this skill, to become more attuned to your intuitive voice. And the best part? You're not alone on this journey. Anyone can unlock the potential of intuition, and with practice, it can get better.

Intuition and AI

Artificial intelligence refers to the ability of a computer system to perform duties, including making decisions, that seem like they're being done with intelligence. Rather than following strict, pre-programmed rules as a traditional computer program would, AI learns from existing data and makes a choice or takes an action based on these prior learnings. AIs can also be generative, producing things like artworks, texts, ideas and decisions.

As far as we know, AIs currently in use do not have consciousness, hence we can consider them as unconsciously acting and deciding agents. However, because we don't yet have a scientific test for consciousness, there is no way to actually test whether an AI is conscious of anything.

Once an AI has learnt information from prior datasets, how does it 'know' what to do and what not to do? The answer is that it knows from linking its learnt information with either good or bad outcomes. In other words, unconscious learning that predicts good or bad outcomes. Sound familiar?

Both AI and intuition involve unconscious learning and the ability to aid and/or to make decisions. As we've seen in the chapter on the fifth rule of SMILE, they both inherit biases from the data they're trained on. If you feed yourself biased or untrue data, your intuition will learn it and become biased or misaligned. AIs will likewise become biased if they are trained on biased datasets. Also, current AIs will easily become biased, misaligned, or simply not function when they are taken out of their intended environment. This applies also in terms of the context of the datasets the AI trained with; for example, writing text versus driving a car. In other words, AIs are context specific, just like your intuition. One possible application of AI to intuition is for the tracking of and learning from our decisions: 'Joel, are you sure you want to have another coffee? Last time you had three coffees in one day you regretted it.' Or perhaps: 'Joel, I just noticed that when you met that person your physiology spiked, was it something they said?' There are a hundred other moments in the day when an AI assistant could point out a bias in our human decisions. As previously mentioned, it's likely that AIs will have their own biases, but this doesn't mean we couldn't use them to track and point out our own. In a very real sense, this is akin to outsourcing *biological intuition*, or at least some of it, to *artificial intuition*.

Another interesting application of AI might be for people with an addiction. As we've seen, those with an addiction tend to make more impulsive, short-term decisions; an AI assistant could be useful with this. When it notices someone making short-term-focused, impulsive choices, the AI could point this out, emphasising the value in more long-term-focused decisions.

The big question is, what would happen to our biological intuition if we were to continuously outsource it? Would it wither and shrink? Because the data we have on intuition and learning does suggest that it requires use for maintenance, without practice, biological intuition would become weaker and weaker.

This is a process called enfeeblement. It would likely mirror the case of phone numbers. I, for one, stopped remembering them a while back, and don't even think of phone numbers anymore; my phone just connects me to people, and the numbers work behind the scenes.

There are many pros and cons of outsourcing our thinking and learning to AIs. If the right advice from psychologists and neuroscientists is embedded into the development of such AI, it seems logical that we can make technology that is human-based and helps us without hurting us, without the enfeeblement of human intuition. But that is a *big* if. These systems could be designed not only to help us make our decisions, but also to help us learn to make better decisions.

Where to from here?

I set out to do something ambitious in this book: to introduce a new theory of intuition and to tether it to a real-world, hands-on toolkit driven by the engine of science, all aimed at improving our decision-making. And aimed always at charting a safe course for you to replicate the successes of people like Jason, Jon, Tom and Jasmine. You don't need to be a pilot or a mountaineer or anything extreme to benefit from developing your intuition. The true value for most of us will lie in the multitude of smaller, everyday decisions, whose true value will become evident as the effects of better decision-making accumulate across the days and weeks.

As we have seen, the science of intuition is in its infancy, and the definition I put forward in this book is a working definition, one that is open to new additions and adjustments – a flexible blueprint that can evolve with new discoveries.

To recap: Intuition is the learnt, productive use of unconscious information to improve decisions or actions.

This definition is optimised to be the most useful. It seeks to be a catalyst to propel scientific discourse, and to transcend misgivings and disagreements about the nature of intuition. Intuition, it becomes clear upon close scrutiny, isn't monochromatic. Its hues come alive as we dive deep, enabling us to recognise when it's our ally and when it may lead us astray – let's not forget to SMILE.

The Practice of Intuition

It's possible that some parts of this book may stir controversy. That's fine by me. The intent here is to electrify the discussion around intuition, provoke deeper thinking, encourage more exploration, foster broader use, and thus grow our understanding. The downstream effect of this can only be better decision-making and actions for all of us.

I foresee a future where intuition assumes a leading role in decision-making in everyone from the top tiers of leadership to sports professionals, from policymakers to new parents. By creating a clear, science-based theory of intuition, we pave the way for its mainstream acceptance as a decision-making tool.

Intuition isn't mystical, it's a skill, something we already understand with science. It's a practice we can cultivate, hone, and should feel free to openly discuss. I invite you to commence your intuitive journey, if you haven't already, taking it one decision at a time. What are you waiting for?

Acknowledgements

First and foremost, I am deeply grateful to those who generously shared their personal stories with me, especially Jason the pilot and Jon Muir. Their narratives have added richness and depth to this book.

I owe a significant debt to the many researchers, authors and scientists, both named and unnamed in this book. Their ground-breaking discoveries and theories about the human mind and brain have greatly enriched my understanding and informed the advice I offer throughout these pages. I thank Adam Alter for his many acts of generosity in support of this book, and Ed Catmul for his thorough feedback and support.

A special note of appreciation goes to the dedicated team at Simon & Schuster Australia. I thank Dan Ruffino and Ben Ball for their unwavering support. I am particularly grateful to Meredith Rose for her incisive suggestions, keen edits, and invaluable writing lessons. A heartfelt thank you to the UK publications team at Welbeck, Beth Bishop and Matt Tomlinson. To Gabby Oberman and Anna O'Grady for their valuable work in PR. Andrew Kirky deserves a heartfelt thank you for his early

Acknowledgements

endorsement and networking assistance, for highlighting the most pertinent knowledge and for training me to be on the lookout for the value added for the reader. Julie Gibbs' belief in the power and significance of intuition as a decision-making tool, as well as her recognition of the value in my theories, has been immensely motivating; I thank her also for connecting me with Dan. I thank Emma Norris for valuable proofreading.

I cannot overstate my gratitude to my agent, Rach Crawford. From our initial interactions, she expertly guided me through the intricate world of publications and publishers. Her insights into the book proposal and overarching strategies have been indispensable, and continue to be so, and also to Kate Johnson for representation support in the UK.

I extend my warm thanks to my entire research lab at the University of New South Wales, both past and present members, and in fact the whole of the university. Dr Galang Lufityanto, during his PhD tenure in the lab, contributed immensely to our understanding and measurement of intuition. My academic journey has been enriched by mentors such as Professors Colin Clifford, Randolph Blake, Frank Tong and Duje Tadin. I thank the Australian Research Council, and the National Health and Medical Research Council of Australia for their generous fellowships over the years, which have enabled me to focus on research and thinking. I thank Gavan McNally for our regular chats about the state of the world, research and academia. I thank my small group of school friends, you know who you are, for keeping me

Acknowledgements

balanced and sane. I thank Walter Isaacson and his book *Steve Jobs* (Abacus, 2011), for its detailed reports of Jobs and his use and misuse of intuition.

The many journalists who have engaged with my theories over the years have offered me fresh perspectives by posing thought-provoking questions about the practical application of my discoveries and theory of intuition. Their insights have been invaluable in shaping things.

The family remains the cornerstone of all endeavours. My mother's steadfast support has been a guiding light. My children, albeit unknowingly, have endured my occasional absent-mindedness with grace, providing inspiration when I needed it most. The ability of my wife – my biggest critic and advocate – to transform the more esoteric sections of this book (read academic) into relatable and engaging content has been invaluable. If you've found this book at all engaging and captivating, a lion's share of that credit goes to her: thank you.